Advanced Introd

Elgar Advanced Introductions are stimulating and thoughtful introductions to major fields in the social sciences, business and law, expertly written by the world's leading scholars. Designed to be accessible yet rigorous, they offer concise and lucid surveys of the substantive and policy issues associated with discrete subject areas.

The aims of the series are two-fold: to pinpoint essential principles of a particular field, and to offer insights that stimulate critical thinking. By distilling the vast and often technical corpus of information on the subject into a concise and meaningful form, the books serve as accessible introductions for undergraduate and graduate students coming to the subject for the first time. Importantly, they also develop well-informed, nuanced critiques of the field that will challenge and extend the understanding of advanced students, scholars and policy-makers.

For a full list of titles in the series please see the back of the book. Recent titles in the series include:

The Sociology of the Self
Shanyang Zhao

Artificial Intelligence in Healthcare
Tom Davenport, John Glaser and Elizabeth Gardner

Central Banks and Monetary Policy
Jakob de Haan and Christiaan Pattipeilohy

Megaprojects
Nathalie Drouin and Rodney Turner

Social Capital
Karen S. Cook

Elections and Voting
Ian McAllister

Youth Studies
Howard Williamson and James E. Côté

Private Equity
Paul A. Gompers and Steven N. Kaplan

Digital Marketing
Utpal Dholakia

Water Economics and Policy
Ariel Dinar

Disaster Risk Reduction
Douglas Paton

Advanced Introduction to

Disaster Risk Reduction

DOUGLAS PATON

Professor, College of Health and Human Sciences, Charles Darwin University, Darwin, Northern Territory, and Adjunct Professor, Faculty of Health, University of Canberra, Canberra, ACT, Australia

Elgar Advanced Introductions

Cheltenham, UK • Northampton, MA, USA

© Douglas Paton 2022

All rights reserved. No part of this publication may be reproduced, stored in a retrieval system or transmitted in any form or by any means, electronic, mechanical or photocopying, recording, or otherwise without the prior permission of the publisher.

Published by
Edward Elgar Publishing Limited
The Lypiatts
15 Lansdown Road
Cheltenham
Glos GL50 2JA
UK

Edward Elgar Publishing, Inc.
William Pratt House
9 Dewey Court
Northampton
Massachusetts 01060
USA

A catalogue record for this book
is available from the British Library

Library of Congress Control Number: 2022944612

ISBN 978 1 80392 045 0 (cased)
ISBN 978 1 80392 047 4 (paperback)
ISBN 978 1 80392 046 7 (eBook)

Printed and bound in Great Britain by TJ Books Limited, Padstow, Cornwall

I would like to dedicate this book to my Mother and Father. They nurtured my love of learning, showed me how to apply knowledge with integrity, humility and compassion, and instilled in me the importance of never stopping asking questions. For this, I am eternally grateful.

Contents

List of figures x
About the author xi
Acknowledgement xii
List of abbreviations xiii

1 Introduction to disaster risk reduction 1
   1.1 The need for disaster risk reduction 1
   1.2 The Sendai framework for disaster risk reduction 2
   1.3 The SFDRR priorities 3
   1.4 SFDRR guiding principles 5
   1.5 Putting DRR into action 7

2 Understanding disaster risk 10
   2.1 Disaster risk 10
   2.2 The environmental context of DRR: the hazardscape 11
   2.3 Hazard characteristics and behaviours 12
   2.4 Conclusion 20

3 Anticipation 22
   3.1 Why is anticipation important? 22
   3.2 Anticipating risk and preparedness needs 23
   3.3 Temporal influences on cost and benefit judgements 31

4 Preparedness 33
   4.1 Preparing 33
   4.2 What does being prepared mean? 34
   4.3 Preparedness and resilience 34

	4.4	Preparedness information	37
	4.5	Understanding and facilitating preparedness	38
	4.6	Preparedness theories	42
	4.7	Preparedness and governance	51
	4.8	Conclusion	54
5		**DRR in international contexts: cross-cultural issues**	**56**
	5.1	DRR in international contexts	56
	5.2	Cultural influences on DRR processes and outcomes	57
	5.3	Cultural diversity bringing a collective strength for all humanity	60
	5.4	Cross-cultural DRR	62
	5.5	Assessing the cross-cultural applicability of the CET	63
	5.6	Socio-cultural-environmental relationships and DRR	68
	5.7	The community consciousness model	69
	5.8	Conclusion	71
6		**DRR in response and recovery settings**	**74**
	6.1	DRR in response and recovery settings	74
	6.2	Applying coping and adaptation constructs to recovery	76
	6.3	Survivor perspectives on response and recovery demands and adaptive capacities	78
	6.4	Adaptive capacities in Christchurch	84
	6.5	Adaptive capacities in Taiwan	86
	6.6	Conclusion	87
7		**Assessing the effectiveness of DRR: cost–benefit and evaluation perspectives**	**89**
	7.1	Assessing the effectiveness of DRR	89
	7.2	Cost–benefit analyses and DRR	90
	7.3	Evaluation: process, content, outcome and context issues	92
	7.4	A developmental approach to process and content evaluation	93
	7.5	Outcome evaluation studies	101

	7.6	Evaluation in recovery and rebuilding settings	110
	7.7	Conclusion	112
8	**Transformative learning, capacity development and building back better**		**113**
	8.1	Learning from disasters	113
	8.2	Experience and inaction	114
	8.3	Capacity development: repurposing, emergent and transformative processes	115
	8.4	Transformative learning	121
	8.5	Modelling transformative DRR learning	124
	8.6	Governance and transformative DRR learning	125
	8.7	Environmental context: city, place and natural settings	129
	8.8	Cultural dimensions, place and DRR	131
	8.9	Socio-environmental relationships and co-existence	133
	8.10	Pulling the transformative DRR learning threads together	134
	8.11	Conclusion	135
9	**Conclusions and future issues**		**136**
	9.1	Knowing DRR for the first time	136
	9.2	A socio-cultural-environmental framework	137
	9.3	Future preparedness issues	139
	9.4	Community development and DRR	140
	9.5	Transformative learning	142
	9.6	Transdisciplinary strategies	144
	9.7	Adaptive governance	145
	9.8	CBA and evaluation	147
	9.9	Organizational continuity planning	148
	9.10	Learning and collaboration in international settings	149
	9.11	DRR in Indigenous populations	149
	9.12	Conclusion	150

References 151
Index 166

Figures

4.1	A matrix model of person, community and societal preparedness predictors	53
5.1	A comparative analysis of earthquake and volcanic research using the CET in countries situated at different points on Hofstede's (2001) IC dimension preparedness predictors	65
5.2	The relationship between culture-general and culture-specific processes in Taiwan, Japan and Indonesia	66
5.3	The community consciousness model	70
6.1	A summary of the interdependent influences of person, family, community and societal adaptive capacities on individual and collective response to adaptive demands experienced following the 2011 Christchurch earthquake	81
6.2	A summary of the interdependent influences of person, family, community and societal adaptive capacities on individual and collective response to adaptive demands experienced following the 921 (Chi Chi) earthquake in Taiwan in 1999	83
7.1	The Shakeout Drill presented as an evaluation model	107
7.2	The relationship between the CET, community engagement strategies, and changes in preparedness in BRN and control communities	110
7.3	A tentative process, content and outcome framework applying QoL to evaluating recovery and rebuilding	112
8.1	Summary of the DRR transformative learning process	135
9.1	The respective contributions of life experience/ community development and risk management variables to DRR	141

About the author

Douglas is a Professor at Charles Darwin University, Australia, an Adjunct Professor at the University of Canberra, a Senior Research Fellow at the Bandung Resilience Development Initiative (Indonesia), and a Research Fellow at the Joint Centre for Disaster Research (New Zealand). In 2005–06, he was the Australian delegate to the UNESCO Education for Natural Disaster Preparedness in the Asia-Pacific. He was a member of the UNISDR (now UNDRR) RIA sub-committee (2012–16) and served on the Psychosocial Advisory Committee for the Christchurch earthquake (2011–13). In 2014 his role as a Technical Advisor on Risk Communication with the WHO helped develop the community engagement program for the Ebola response program in Sierra Leone. His research develops and tests theories of adaptive and resilient capacity in communities using all-hazards and cross-cultural approaches. Current work includes developing transformative approaches to community capacity building during disaster recovery and using the visual and performing arts to support disaster recovery. He has served as Editor of the *Australasian Journal of Disaster and Trauma Studies* (Founding Editor), *Disaster Prevention and Management* and the *International Journal of Mass Emergencies and Disasters*. He sits on several journal editorial boards, including the *International Journal of Environmental Research and Public Health*, *Disasters*, and the *International Journal of Disaster Risk Reduction*.

Acknowledgement

I would like to express my special thanks to Petra Buergelt for providing valuable feedback on earlier versions of the manuscript.

Abbreviations

BBB	Build Back Better
BRN	Bushfire-Ready Neighbourhoods
CA	Critical Awareness
CBA	cost–benefit analysis
CERT	Community Emergency Response Team
CET	Community Engagement Theory
DRR	Disaster Risk Reduction
HBM	Health Belief Model
IC	individualism–collectivism
LTO	long-term orientation
MF	masculinity–femininity
NERT	Neighbourhood Emergency Response Team
NSW	New South Wales
PADM	Protective Action Decision Model
PD	power distance
PMT	Protection Motivation Theory
PrE	Person-relative-to-Event Theory
QoL	Quality of Life
SFDRR	Sendai Framework for Disaster Risk Reduction
TPB	Theory of Planned Behaviour
UA	uncertainty avoidance
UNDRR	United Nations Office for Disaster Risk Reduction

1 Introduction to disaster risk reduction

> The superior man, when resting in safety, does not forget that danger may come. When in a state of security, he does not forget the possibility of ruin. When all is orderly, he does not forget that disorder may come. Thus, his person is not endangered, and his States and all their clans are preserved.
>
> Confucius (551–479 BCE)

1.1 The need for disaster risk reduction

Throughout human history, societies have emerged and flourished in places which afford their citizens opportunities to benefit from, enjoy and utilize the physical, economic and aesthetic amenities offered by their natural environment. For example, natural harbours and coastal scenery can derive from local seismic activity, commerce flourishes from using rivers as trade routes, volcanic ashfall can create productive soils, and forests afford amenity and employment benefits to those living nearby. However, periodically the natural sources of these benefits turn hazardous. In doing so, they pose significant threats to societies, citizens and the institutions and infrastructure that support people's lives and livelihoods.

Consequently, if citizens and societies are to continue to enjoy the benefits they derive from their relationship with their natural environment, they must develop ways to anticipate, cope with, respond to, adapt to, recover from, and learn from their periodic encounters with these hazardous circumstances. Contemporary approaches to facilitating these capabilities are collectively encompassed by the Disaster Risk Reduction (DRR) concept.

The opening quote from Confucius suggests that, in essence, DRR is not an entirely new idea; attempts to encourage people to act to protect themselves from danger, ruin and disorder have a history going back some 2,500 years. While not embodying all the tenets of contemporary DRR, Confucius raises some pertinent issues. For example, he emphasizes responsibility as being vested in the person rather than relying on external assistance, highlights the importance of anticipating one's risk (does not forget that danger, ruin or disorder may come), especially during periods of hazard quiescence, and describes how such vigilance culminates not only in protecting oneself, but also protecting one's family and the wider community. These goals align well with those of contemporary DRR. Contemporary DRR is not, however, without its challenges.

Despite consistent attempts to encourage their implementation, people have proven reticence to adopting DRR recommendations. It is to be hoped that the emergence of the Sendai Framework for Disaster Risk Reduction (SFDRR), and its creation of a systematic foundation for DRR, will mark a turning point in DRR becoming established. This book seeks to contribute to this process by providing evidence-informed insights into understanding people's (individual and collective) reticence to engage with DRR processes and identifying how to reverse this trend and facilitate people's active participation in DRR in ways that support realizing the SFA goals. First, these goals need to be introduced.

1.2 The Sendai framework for disaster risk reduction

The systematic description of what comprehensive DRR should encompass took a significant step forward with the emergence of the Sendai Framework for Disaster Risk Reduction (SFDRR) (Munene et al., 2018). The major goals of the SFDRR are twofold. The first is to facilitate substantial reductions in disaster risk and the loss of lives, livelihoods, and well-being outcomes that disasters create. The second is to develop an all of society approach to protecting the economic, physical, social, cultural and environmental assets of persons, businesses, communities and countries. These are encapsulated in the SFDRR Priorities and Guiding Principles.

1.3 The SFDRR priorities

The SFDRR Priority 1, Understanding Disaster Risk, argues that DRR research and practice must derive from understanding environmental contributions to disaster risk. It also calls for researching how this knowledge informs understanding how and why societies, citizens and their personal and societal assets are exposed to potential losses, how this exposure creates sources of vulnerability, and the implications this has for how societies and citizens can proactively develop their DRR capability.

Priority 1 focuses on data, information, knowledge, evidence, and disaster research, but emphasizes that the outcomes of these processes must be tailored to suit the needs of their intended users. The all-of-society focus advocated by the SFDRR draws attention to the need for mechanisms to manage DRR at national, regional and local levels. This introduces the role of disaster risk governance, a topic that is the subject of Priority 2.

The SFDRR advocates for DRR having a strong societal foundation through its Priority 2, strengthening disaster risk governance to manage disaster risk. Governance describes the policies, procedures, practices, norms and relationships developed and institutionalized by agencies and organizations to accomplish societal DRR (e.g., Berkes, 2021). Governance facilitates how the SFDRR's reduction, preparedness, response, recovery and rebuilding goals are implemented in ways that reduce disaster impacts, losses and consequences, provide strategies to enable coping and adaptation, and support learning from events in ways that culminate in the incorporation of lessons learnt into future DRR capacities and capabilities.

Priority 2 describes how disaster risk governance must encompass global, national, regional and local levels of analysis in ways that support the development of coherent (and complementary) national and local frameworks of laws, regulations and public policies. The latter, in turn, lays the foundation for defining stakeholder (individual and collective) roles and responsibilities in ways that guide, encourage and incentivize DRR action across all sectors of society.

Rather than allocating a specific chapter to it, the role of governance is incorporated in the discussions of how disaster risk is understood (Chapter 2), its implications for managing diversity in DRR outcomes

(Chapters 3 and 4), its influence on recovery planning and intervention (Chapter 5), its role in managing international collaboration (Chapter 6), and its influence on capacity development (Chapter 8). These discussions highlight how the SFDRR goals cannot be realized without citizens and societies incurring costs. These can be considerable when large-scale mitigation projects (e.g., constructing levees) are undertaken. The latter issue introduces the SFDRR Priority 3; the need for DRR processes to provide a return on citizen and societal investments in DRR.

Priority 3 advocates that public and private investment in DRR through structural (e.g., land-use planning, building codes) and non-structural (e.g., preparedness) actions should protect lives and assets, reduce or prevent losses, enable effective recovery and rehabilitation, and enhance the economic, social, health and cultural resilience of persons, communities, countries, their assets and their environment. It also argues that the DRR initiatives developed to achieve these outcomes should be cost-effective and result in a return on investment for societies and citizens. However, achieving such goals in practice are complicated by the need for cost–benefit analyses to encompass tangible (e.g., constructing levees) and intangible costs. An important representative of the latter category is the focus of Priority 4, preparedness.

The SFDRR Priority 4 advocates for DRR to include strategies to proactively enhance disaster preparedness in ways that reduce people's risk and enable their effective response to hazard events. There exist several ways to reduce risk, including through land-use planning (e.g., regarding avoiding development in earthquake-prone areas), engineering solutions (e.g., building earthquake-resilient buildings; engineering soils to reduce risk), warning systems and preparedness (Fraser et al., 2016).

The United Nations Office for Disaster Risk Reduction (UNDRR, 2016) defines preparedness as the *knowledge and capacities developed* by governments, response and recovery organizations, communities and individuals to effectively *anticipate*, *respond to* and *recover from* the impacts of *likely*, *imminent* or *current* disasters. Recognition of people's pre-event reticence to prepare for disaster prompted the inclusion of an important component to Priority 4 – the Build Back Better (BBB) construct. An important contribution of the latter is it appreciating that the magnitude of disaster-related losses originates, at least partly, from inadequate pre-event DRR planning and capability. Priority 4 proposes that including a developmental focus in recovery and rebuilding initiatives can reverse this trend by capitalizing on people's disaster experience and using it as a springboard for increasing citizen and societal capacity to deal more effectively with future disasters.

Priority 4 also identifies the urgent need to accommodate women and persons with disabilities in planning and development, and ensure that DRR investment goals (Priority 3) include promoting gender-equitable and universally accessible approaches to DRR and to response, recovery and reconstruction phases. How these Priorities can be enacted is addressed in a series of Guiding Principles.

1.4 SFDRR guiding principles

The Guiding Principles proposed by the SFDRR place primary responsibility for DRR on (nation) states. However, the SFDRR also advocates for socially inclusive approaches to DRR, including through cooperation across national, regional and local levels of analysis. The shared responsibility, multi-stakeholder cooperative and collaborative issues this introduces for all aspects of DRR will be discussed throughout this volume.

The *shared responsibility* principle advocates that central government and national authorities, sectors and other stakeholders play, as far as national circumstances allow, complementary roles in DRR. This extends into the realm of social justice to include sharing responsibility for safeguarding, promoting and developing all human rights, including the right to development. This draws attention to a need to consider procedural and distributive justice in how shared responsibility principles and strategies are conceptualized, implemented and evaluated when operating at a whole-of-society level.

The SFDRR Principles depict full societal engagement in terms of developing relationships between national/country institutions and their associated executive and legislative processes in ways that enable the emergence of complementary plans and actions at state/regional and local levels of government. Pivotal to operationalizing this principle concerns how such arrangements can empower local authorities to develop local incentives, resourcing, planning and the creation of (shared or distributed and localized) decision-making responsibilities in ways that encompass the needs and goals of the diverse (geographically, socially, culturally, etc.) communities within their jurisdictions. The latter is essential if DRR practices are to further the SFDRR goal of developing coherent and socially inclusive stakeholder capacities, local development policies, plans, and practices and mapping them onto local risk profiles (derived from multi-hazard/all-hazards risk assessments) to facilitate locally relevant DRR capabilities. The implications of these issues will be discussed throughout the book but primarily in relation to governance, preparedness and BBB or capacity development strategies. These issues are discussed in relation to their potential to contribute to cost-effective DRR through better locally targeted investment that entails applying an evolving, iterative and proactive approach to managing future risk by leveraging disaster experience as a catalyst for developing sustainable future capacities and capabilities (see Chapter 8).

The final SFDRR Principle reiterates the importance of thinking globally. Thinking global has additional implications for understanding disaster risk. For example, climate change processes and large-scale hazard (e.g., fires in Indonesia and volcanic eruptions in South America and Iceland whose impacts extended across international borders) events can create international consequences (both in regard to the physical sources of hazards and the political contexts in which hazard events occur) that need to be managed as such. Thinking globally raises some additional issues.

One issue concerns a need to develop high-quality global partnerships and levels of international cooperation that can contribute to the effective management of transnational impacts. Including an international dimension in DRR also draws attention to a need to consider how relationships between the developed and the developing world should be accommodated in DRR needs and priorities. A second issue relates to the cultural diversity that adopting a global perspective introduces into DRR and the implications this has for collaboration and shared learning across national borders. While providing a comprehensive framework for action, the complex interdependences involved (e.g., multi-level, multi-stakeholder and global engagement in activities ranging from preparedness to recovery to post-event learning) makes it important to reflect on what this means for DRR in action.

1.5 Putting DRR into action

The SFDRR Framework includes several pivotal ideas. These include the importance of adopting a whole-of-society approach in which governments, public and private sector agencies, NGOs, and citizens and social networks share responsibility for developing, implementing and evolving DRR. Another is the inclusion of an all-hazards approach to facilitate developing a comprehensive DRR capability. One valuable contribution derives from acknowledging that disaster losses derive partly from inadequacies in prior planning, engagement and implementation. The consequent inclusion of the BBB construct makes an important contribution to conceptualize DRR as an iterative, evolving process in which multiple stakeholders will play crucial and interdependent (shared responsibility) roles. While all of these are amenable to implementation, a challenge remains regarding their integration into a coherent framework.

A major impediment to enacting the SFDRR is a lack of a conceptual framework capable of integrating the priorities and the absence of consensus regarding how the SFDRR Priorities and Principles can be operationalized and implemented (Munene et al., 2018). As a first step in creating a conceptual foundation, this book situates DRR in a socio-cultural-environmental framework (Buergelt & Paton, 2014; Karpouzoglou et al., 2016; Paton, Buergelt & Campbell, 2015; Twigg, 2015).

This socio-cultural-environmental framework is used to demonstrate how social, social diversity (beliefs, demographics, etc.), cross-cultural and multicultural perspectives (with these collectively describing *socio-cultural-* inputs) interact with all-hazard or multi-hazard viewpoints on disaster risk (which adds an *-environmental* input) make interdependent contributions to the theories, policies and practices that can contribute to realizing the SFA's goals. This is supported by discussing how evidence-informed strategies, situated with this framework, can be developed and enacted in ways that facilitate the proactive and sustainable development of people's individual and collective capacities to anticipate, prepare for, respond to, recover from, and learn from the impacts of environmental hazard activity that will only increase in the future.

This socio-cultural-environmental (including both urban and natural settings) framework is referred to throughout the text to illustrate how the issues canvassed in the Priorities play complementary roles in comprehensive DRR programs. This starts with demonstrating how disaster risk can only be fully understood from a socio-cultural-environmental perspective. This is introduced in Chapter 2

Chapter 2 focuses specifically on Priority 1, Understanding Disaster Risk. In doing so, it illustrates how DRR can be conceptualized as socio-cultural-environmental phenomena in which people rely on interpretive capacities and social-cognitive biases to interpret hazard characteristics and behaviours in ways that can undermine their motivation to engage in DRR activities. The discussion of how interpretive and social-cognitive biases influence DRR outcomes continues in Chapter 3.

Chapter 3 singles out anticipation as a construct that merits greater consideration in DRR than has hitherto been the case. This position is justified by discussing how several interpretive, emotional and social-cognitive biases can undermine people's interest in or ability to anticipate their future risk. These constraints can culminate in people disengaging from DRR processes. This chapter discusses the steps that can be taken to counter these constraints and enable people's ability to anticipate their need for preparing.

Chapter 4 introduces preparedness. It discusses why people's initial interest in preparing focuses primarily on low-cost survival preparedness, and outlines strategies that enable their developing more comprehensive levels of preparedness. This includes discussion of the governance implications of developing and maintaining preparedness. The SFDRR's inclusion of a global perspective introduces a need to consider the implications of cultural diversity for DRR beliefs, practices and processes, including governance. This issue is introduced next.

Chapter 5 recognizes that while the sources of hazards such as earthquakes and floods occur worldwide, the socio-cultural characteristics of the peoples they affect differ from country to country. Chapter 5 draws on comparative cross-cultural studies of preparedness to explore opportunities to use DRR processes to support international learning and collaboration. It also introduces how knowledge of culture-specific processes must be accommodated in developing and applying practical DRR strategies.

Next, in Chapter 6, attention turns to how, when faced with diverse and evolving disaster response and recovery challenges, people demonstrate an ability to prepare in situ in ways that enable their being to cope with and adapt to recovery and rebuilding demands that can span months or years. This includes how recovery draws on several societal and community resources. The level of personal and societal investment involved in developing relevant capabilities draws attention to this issue discussed next; can these activities demonstrate a return on this investment?

Chapter 7 opens with a brief summary of work on cost–benefit analysis and the challenges that arise when applying this process to the social and environmental facets of DRR. In response to these challenges, Chapter 7 proposes turning to process, content and outcome evaluation strategies to provide a foundation for assessing the effectiveness of social and environmental interventions. The chapter discusses four approaches to evaluation; a comprehensive developmental evaluation model, an exercise (Shakeout Drill) evaluation, a theory and community development-based evaluation model, and an approach that explores how Quality of Life measures can be used to evaluate the effectiveness of recovery and rebuilding initiatives.

Chapter 8 ends the substantive chapters by introducing the challenges associated with transitioning recovery and rebuilding activities into Build Back Better or capacity development outcomes. This discussion focuses on how disaster experience can act as a catalyst for transformative learning and generate substantive shifts in people's DRR beliefs, knowledge and capacities, including how transformative learning outcomes can be consolidated in the socio-cultural-environmental fabric of social and societal life. The book closes in Chapter 9 with a discussion of progress to date in realizing the vision outlined in the SFDRR and outlines several areas for future work. The journey begins with Chapter 2's introduction to understanding disaster risk.

2 Understanding disaster risk

2.1 Disaster risk

> Civilisation exists by geological consent, subject to change without notice.
> (Will Durant (1885–1981))

This quotation aptly encapsulates the importance of DRR. If the activity of the geological and other natural processes Durant alludes to directly impact on societies and their citizens, they can potentially create devastating consequences for civilization, and can do so with no or little warning. However, the risk posed by such occurrences and the significance of the consequences societies and citizens could experience is not a fait accompli. There is much people can do to reduce their risk and increase their capacity to respond and recover should disaster strike. The development of these capacities starts with understanding the sources of the risks that create the consequences that must be reduced and, if necessary, responded to and recovered from. Chapter 1 introduced how DRR is situated in a socio-cultural-environmental framework.

This chapter focuses on the *environmental* component of this relationship and introduces the natural processes that represent the sources of hazard consequences that societies and citizens must contend with. In doing so, it draws attention to how the socio-cultural side of this relationship introduces how, for example, people's beliefs and the emotional reactions elicited by considering hazard issues influence people's interpretation of risk information in ways that determine whether and how people use or do not use risk data.

The importance of understanding natural environmental sources of risk is described in the SFDRR Priority 1, *Understanding Disaster Risk*. This priority describes the importance of relating DRR beliefs, plans and actions with knowledge of environmental contributions to disaster risk. The quality of this relationship informs the development of several DRR processes, from where infrastructure development will occur, to the structural mitigation measures required, to land-use and urban planning decisions, to the governance (national, regional, local) policies and practices that shape DRR strategies and outcomes, to how risk information is communicated to people. All of these processes are built on understanding the environmental context in which DRR is situated.

2.2 The environmental context of DRR: the hazardscape

The task of *understanding disaster risk* (Priority 1) starts with understanding the hazards present within a given environment. The mix of hazard sources, characteristics and behaviours within a given environment, geographical area or jurisdiction describes its hazardscape. A knowledge of hazardscapes describes the sources of risk within a given area or jurisdiction and informs the development of DRR beliefs and processes (e.g., governance, preparedness). An important consideration here is appreciating that, within a given jurisdiction, risk can emanate from multiple sources as well as from secondary and cascading hazard activity.

For example, citizens in certain areas of California face risk from seismic, volcanic and wildfire sources and residents in the mountainous areas of Taiwan must manage risk from diverse geological (e.g., earthquake) and meteorological (e.g., typhoon) hazards. In both locations, secondary and cascading hazards (e.g., landslides triggered by seismic activity or from the action of intense rainfall) become additional sources of risk. These examples highlight the value of conceptualizing DRR from an *all-hazards* perspective (UNDRR, 2015). The process of understanding disaster risk is further complicated by the potential for hazardscapes to change over time.

The data and information that contribute to contemporary *understanding of disaster risk* represents a body of systematic research stretching back over a few hundred years. As this research continues, the understanding of natural process and the sources of risk they create is continually evolving and is doing so in some interesting ways.

For example, Cochran et al. (2004) found a relationship between tidal strength and the occurrence of shallow thrust faults, with seismic activity being affected by changes in the weight of water pressing on seismically active areas. Consequently, sea level rises associated with climate change processes could increase levels of seismicity. Another example of how environmental risk research can influence DRR thinking and actions comes from work on Auckland's (New Zealand) volcanic field. Shane et al. (2013) found that eruptions could persist for decades, rather than the weeks or months upon which previous risk management thinking was based, with these findings introducing significant changes to the required planning, relocation and preparedness strategies. Climate change is also changing hazardscapes. For example, Chen and Chen (2015) discuss how climate change has resulted in a need to add wildfire to the list of hazards present in tropical countries like Taiwan.

These examples introduce how research is providing some novel insights into how hazardscapes are evolving. As this and other sources of disaster risk knowledge develop, so too does the need to consider how the risk posed by these hazardous circumstances is communicated and understood, including its implications for people's understanding of their relationship with their environment, the challenges they face in making sense of their risk, and what they need to do to enable their co-existence with their dynamic hazardous environment. An important issue here is that knowing of changes in hazardscapes per se is only one aspect of understanding disaster risk. The other derives from understanding the hazard characteristics and behaviours that derive from their hazardscapes, and how people interpret them. The latter is a major influence on both people's understanding of their risk and whether or not they use risk data to inform their DRR decisions and actions.

2.3 Hazard characteristics and behaviours

To develop socio-cultural-environmental co-existence strategies, people need to know two things about their potentially hazardous natural environmental circumstances. The first concerns the natural processes (e.g., earthquakes, wildfires) present within the environments they inhabit or rely on and from which the risks that need to be managed emanate. The second covers people's knowledge of hazard characteristics and behaviours that determine the DRR actions they need to implement.

The sources of loss, damage and harm people experience derive from *hazard characteristics* such as ground shaking from seismic activity or ember attack from wildfires (Gregg & Houghton, 2006). These characteristics are present every time hazard activity occurs. They represent the sources of the consequences people must anticipate and prepare for. While hazard characteristics are generally perceived as environmental threats, this is not always the case. For example, Gregg et al. (2008) discussed how, for Indigenous Hawai'ians, lava flow hazards are revered rather than being perceived as sources of threat (see Chapter 5). Hence, socio-cultural beliefs can influence perception of risk information. Returning to discussing disaster risk, it is a second phenomenon, hazard behaviours (e.g., return and precursory periods, speed of onset, intensity, duration) that make the most significant contribution to understanding of disaster risk.

Thus, while characteristics such as ground shaking or ember attack will accompany every earthquake or wildfire that occurs, each will differ regarding, for example, the intensity of ground shaking or volume of embers that accompany each event. These dynamic *hazard behaviours* are crucial to people's understanding of their risk and its DRR implications. However, complexities in how sources of risk are classified, described and communicated introduce several challenges to people's ability to interpret and make sense of disaster risk data, with the outcomes of these interpretive processes determining how they use, or do not use, these data to make their risk management choices. Hazard behaviours include, for example, return and precursory periods, speed of onset, intensity, duration and so on. These are introduced next, starting with the most community cited hazard behaviours, frequency of occurrence.

2.3.1 Frequency of occurrence

The frequency of a natural process describes how often it occurs within a specific time period. Frequency data allows estimates of return periods, and thus approximations of when a future hazard event could occur.

Frequency or return period data exercises significant influences on people's judgements about the urgency of preparing. However, citizens' judgements about return periods are often inconsistent with those used by scientists and risk management professionals. A common source of citizen (and others) misunderstanding derives from return period estimates being calculated by dividing the total number of years over which events have occurred by the total number of events.

For example, if an event occurred 10 times in 200 years, it would have a frequency of 0.05 (i.e., the probability of occurrence in any one year is 0.05 or 5 per cent) or, on average, one event (e.g., an earthquake) every 20 years. The shorthand description of this as a *20-year event* can manifest as people interpreting *1 in 20 years* as prescribing a specific time frame for recurrence (e.g., 20 years hence) rather than an annualized probability (i.e., a 1 in 20 or 5 per cent chance in any given year). This kind of misinterpretation can diminish the likelihood of citizens engaging with DRR processes.

As return periods are pushed out (e.g., 1 in 20 years versus 1 in 50 years), people perceive progressively less urgency to act, lessening their perceived need to prepare or take action to manage their future risk (Paton et al., 2005). Paton and colleagues found that levels of preparedness were relatively high amongst those who interpreted risk as an annualized probability (occurring within the next 12 months), but dropped sharply amongst those who adopted a 50-year event definition that resulted in their believing that the next earthquake would not occur until a more distant future time. Work on message framing to increase the accuracy of return period interpretation has shown promise as a communication strategy capable of helping people develop both realistic risk estimates of their risk and encourage their preparedness (McClure et al., 2009).

While frequency is the hazard behaviour presented most often in public education, frequency data per se does not inform people about how damaging or persistent a given event may be when it occurs. Consequently, a comprehensive understanding of disaster risk requires considering how frequency (i.e., likelihood of occurrence) interacts with data on other hazard behaviours such as intensity and duration. These latter behaviours have important implications for informing people of the consequences they can experience when natural processes (e.g., earthquake, flood) interact with people and human use systems (e.g., buildings, infrastructure). The role of hazard intensity is discussed first.

2.3.2 Intensity

Intensity data provides a measure of rate of impact (e.g., as energy per unit time) and thus affords making estimates of physical (e.g., probable damage) impacts (Gregg & Houghton, 2006). Measures such as the modified Mercalli earthquake scale link intensity estimates with building damage. Knowledge of the range of intensities that can occur informs decisions about mitigation criteria (e.g., the height of a levee) and provide input into planning and cost–benefit analyses (Priority 3). However, at a societal level, it is important to appreciate that planning decisions, and especially those relating to societal mitigation decisions, can include trade-offs that reconcile economic and political considerations against the level of structural mitigation developed (e.g., build to deal with a likely rather than a worst-case, highest intensity event).

Even when mitigation measures are designed to accommodate high-intensity events, other unexpected coincidental events can influence their effectiveness. An example of the latter is how unexpected ground subsidence effectively reduced sea wall height following the Tōhoku earthquake and tsunami in Japan in 2011. Trade-off decisions can create additional social issues.

While, for example, land-use planners and risk managers may understand the rationale for trade-offs in mitigation decision-making, a failure to communicate this to the population at large can result in lay citizens overestimating the level of protection a mitigation measure affords them, reducing their perceived need for personal preparedness. This process, known as risk compensation (or risk homeostasis), can influence people's anticipatory thinking (see Chapter 3). The communication of intensity information should be accommodated in risk communication processes (McClure et al., 2007). Another parameter that contributes to a comprehensive understanding of disaster risk is event duration.

2.3.3 Duration

Natural hazard events can vary in duration from minutes and hours (e.g., individual earthquakes/aftershocks, tornadoes, landslides, avalanches, flash floods) to days/weeks (e.g., some floods, wildfires, earthquake aftershock sequences) to decades (e.g., some volcanic eruptions, drought, earthquake aftershock sequences). Duration need not always be continuous.

For example, earthquake aftershock sequences can be experienced as a series of acute events spread over months or years (Becker et al., 2019). Becker and colleagues allude to the challenges that arise in communicating about aftershock sequence consequences (e.g., progressive damage, a need for self-reliance to occur repeatedly over time). Aftershock sequences have other implications. Even if not causing major damage, aftershock experiences can exact a psychological toll on citizens (Paton et al., 2014).

Misunderstanding or ignoring aftershock data can reduce the likelihood of people appreciating its implications for their safety and reduce their being receptive to their need to develop their preparedness resources in order to cope with and adapt to the challenges encountered in prolonged response periods. An important issue here is that these preparedness needs can differ from those communicated about pre-event preparing for acute events and those required to deal with prolonged duration events. The latter issue is discussed in Chapter 6.

Anticipating the duration of hazard activity that could be experienced has important implications for natural hazard planning and the development of governance policies and practices (e.g., regarding the time frame that societal systems may be disrupted, how long people will have to adapt to circumstances with no or limited access to normal societal services and functions) (Munene et al., 2018). Duration data has additional implications for factors such as evacuation and relocation planning, dealing with loss of utilities (e.g., power, water, sewerage), and communicating the response and preparedness implications of these to relevant stakeholders.

It is also important to appreciate how interaction between factors such as frequency, duration, and intensity parameters influence the demands faced by affected populations. For example, experiencing high-intensity, long-duration events presents more threatening contexts for coping and adaptation, and highlights the importance of including psychological preparedness (see Chapters 4 and 7) and evacuation/relocation preparedness in people's DRR agendas. It is evident here that this is becoming a complex process for people; as the number of risk parameters increases, so too does the demands made on people to find out the values of these parameters where they live and to integrate them into their risk assessment. More on this below. How well people receive and accept these data has important implications for their risk management and preparedness decisions. Duration is not the only temporal characteristic that informs understanding of disaster risk; others include precursory and response times.

2.3.4 Precursory periods and response times

A distinction can be made between the lead time and precursory (or warning) period afforded by a hazard event and the time required for that event to adversely impact a given area (Gregg & Houghton, 2006). The precursory period, or warning time, describes the time interval between detection of precursory activity and the onset of hazard activity (e.g., the time between detecting a potentially tsunamigenic event and confirmation of the formation of a tsunami). Precursory periods can range from minutes (locally generated tsunami) to months or years for some volcanic events.

In contrast, the response time, or the potential speed of onset of hazard activity, describes the time period between the detection of a hazard event (e.g., the formation of tsunami) and the time when the event begins to impact an area (e.g., when a tsunami reaches a coastal settlement). Response times, which can vary from minutes to weeks or longer, delineates the maximum time people have to activate and implement their response plans or, if not prepared or are ill-prepared, the time available to react in ways to minimize the impact of hazard activity for themselves.

Precursory periods and response times are not hazard parameters commonly contemplated in people's DRR decision-making. Additional challenges arise if people elect to use warning information as their cue to prepare rather than as a signal to activate their existing preparedness (Prior & Paton, 2008). This issue is at its most significant for hazards that offer no (e.g., earthquakes) or limited (e.g., local-source tsunami) warnings; using a warning as the catalyst for preparing will leave people with insufficient time to prepare. Conversely, some kinds of hazard events, such as some volcanic eruptions, can have precursory periods that spans weeks, months or years (though this is not the same as being able to predict the actual timing of an eruption). This circumstance can create other preparedness challenges. A key issue here is sustaining people's interest in preparedness if no hazard activity occurs. Long precursory periods can have other consequences.

For example, volcanic events accompanied by long precursory periods may trigger economic fallout. This can include, for example, withdrawal of capital and reduced business investment and activity from areas threatened by an impending hazard event, especially when events are spatially distributed over a large area.

2.3.5 Spatial distribution

Hazards can vary from those that have relatively prescribed spatial distributions (e.g., floods limited by local topographical characteristics) to those having international (e.g., large volcanic eruptions) implications (Mayell, 2002). Hence, the geographical location of a source of hazard activity may provide few clues to the potential distribution of its consequences. Another issue here concerns the uneven distribution of hazard consequences. For example, the distribution of consequences can be affected by geological and topographical features (e.g., variation in topography or rock strata) and by the adoption (or non-adoption) of structural mitigation measures (e.g., building design, retrofitting to safeguard against damage from ground shaking) in designated risk areas.

The spatial distribution of hazard consequences and people's beliefs about this hazard behaviour has other implications for risk management. Support for mitigation is less likely to be forthcoming from those who do not believe they could be affected (e.g., people living hundreds of kilometres from a volcano) even though this remains a possibility (e.g., depending on the intensity of an event). Hence, even if not directly threatened, those living at some distance from a volcano could still be affected by complex distributional factors (Johnston et al., 1999).

The distribution of volcanic (e.g., ashfall) hazard consequences (and similarly for wildfire hazard consequences such as embers and smoke) is affected by how they interact with dynamic meteorological factors such as rain (e.g., heavy rain could concentrate impacts closer to a source) and wind (e.g., strength and direction). Consequently, the spatial distribution and relative intensities of some types of hazards will vary over time and space, with the latter also affecting magnitude impacts. For example, volcanic ash will have less impact on buildings when dry compared with when it is wet from rainfall (which increases weight and acidity).

These dynamic contextual influences add to the challenges citizens face when attempting to assess their risk (Johnston et al., 1999). While events such as earthquakes can occur at any time, some hazard events occur at certain times of year. This introduces temporal distribution as a hazard behaviour.

2.3.6 Temporal distribution

Hazards sourced from meteorological and hydrological processes (e.g., hurricanes, floods) are characterized by their having seasonal patterns of occurrence (though climate change processes are altering traditional patterns of seasonal occurrence). Temporal distribution has implications for people's preparedness thinking.

For example, with seasonal events (e.g., wildfires, floods), people are less likely to engage in anticipatory thinking about this hazard during non-hazard periods (e.g., winter), yet it is often in late winter and early spring that people's preparedness should commence. Increases in seasonal risk will be driven by climate change processes, and with this can come seasonal changes in when people should start preparing and how long they must sustain their preparedness.

This brief discussion introduces how the benefits of situating DRR within a socio-cultural-environmental framework (see Chapter 5 for examples of cultural influences). Knowledge of hazard characteristics and behaviours do have direct implications for hazard scientists and emergency managers involved in areas such as structural mitigation, land-use and warning planning. However, from the perspective of the citizen recipients of these data, it is people's knowledge of and interpretation (or misinterpretation) of hazard characteristics and behaviours that most influences their risk beliefs and DRR (e.g., preparedness) choices. This provides a good introduction to the value of conceptualizing DRR as a socio-(cultural – see Chapter 5 for examples of cultural determinants of risk interpretation and attitudes to mitigation)-environmental co-existence process in which people's actions are influenced by their environmentally situated beliefs, cognitions, resources, capacities, relationships and systems.

2.4 Conclusion

The preceding discussion provided examples of how people's DRR outcomes are influenced by how social (e.g., the risk assessment implications of people's interpretation of hazard characteristics and behaviours, how these interpretations affect their preparedness intentions) and environmental (hazard characteristics and behaviours) factors play interdependent roles in DRR processes. The discussion of hazard behaviours above introduced how societal and citizen DRR outcomes can be affected by people misinterpreting a hazard behaviour (e.g., return period), being unaware of the DRR implications of hazard behaviours (e.g., intensity, duration), or by using behavioural data inappropriately (e.g., relying on warnings as a signal to start preparing). These challenges are compounded in locations where societies and citizens must prepare for the impact of hazards from several sources.

Furthermore, the challenges to DRR emanating from misinterpretation, ignorance of and misuse of hazard behaviour data will grow and evolve as climate change processes alter hazardscapes and introduce new hazards and sources of risk into some areas. The preceding discussion also drew attention to how the dissemination of scientifically oriented (e.g., hazard behaviour) data cannot automatically be assumed to inform how societies and their citizens interpret and understand their disaster risk or make decisions about how they could manage their risk.

It is important to appreciate that the complexity inherent in such interpretive tasks can discourage people from engaging in anticipatory thinking and engaging with DRR processes (Bočkarjova et al., 2009; Gregg & Houghton, 2006; Paton, Kerstholt & Skinner, 2017). Recognition of these problems has stimulated the development of risk communication processes designed to better communicate disaster risk (Becker et al., 2019; McClure et al., 2009).

It follows that if DRR policies, processes and procedures are to realize their goal of facilitating people's understanding of their disaster risk, it is important to understand the personal, social and cultural factors that influence how people interpret dynamic hazard characteristics and behaviours. This knowledge can then inform how to engage people in ways that facilitate their ability to make their DRR choices in ways appropriate for anticipating the risk present in *their* specific locality and circumstances (i.e., their neighbourhood).

If people cannot anticipate their risk, the likelihood of their, for example, supporting societal mitigation strategies or developing their preparedness will diminish. The term *anticipation* is important here.

The UNDRR (2016) definition of preparedness includes a role for *anticipation*. For highly trained scientists and risk management professionals, a fundamental part of their role includes anticipating future risks and their DRR implications. However, it cannot be assumed that lay citizens will find anticipating their risk as straightforward as their scientific and professional counterparts. The potential for the misinterpretation, ignorance of or misuse of disaster risk data introduced above are examples of how interpretive processes can impede citizen risk anticipation. There are others.

Several social-cognitive biases and beliefs and the emotional correlates of anticipating one's risk can be similarly implicated as potential impediments in this context. If, however, the actions of these impediments can be understood, it becomes possible to identify how to circumvent their influence and increase the likelihood that people will start to engage with DRR processes. This issue is discussed in the next chapter.

3 Anticipation

> Our greatest fears lie in anticipation.
> Honoré de Balzac (1799–1850)

3.1 Why is anticipation important?

Large-scale hazard events occur infrequently. Consequently, because people often enter into their DRR deliberations with little or no knowledge of what they could experience when disaster strikes, they first need to anticipate what could occur, what they must plan for and prepare for, and what they will have to respond to. This chapter singles out *anticipation* as an important aspect of people's engagement (or non-engagement) with DRR processes.

While appreciating that the conceptualization of anticipation used here probably differs from what UNDRR (2016) intended, the contents of this chapter will demonstrate why anticipation is a construct that plays a vital role in whether or not people engage in DRR processes. Indeed, the importance of anticipation is mirrored in Balzac's prophetic words. For most people, contemplating disaster is a significant source of fear and anxiety, and one that impedes their ability to anticipate what they will have to contend with when disaster strikes. The chapter introduces the roles emotion and several interpretive biases and beliefs play in impeding people's ability to anticipate their risk. This can even translate into people deciding not to prepare for disaster (Harries, 2008; Paton et al., 2005).

For example, Paton (2007) found that, despite acknowledging their risk (from volcanic hazards), some 58 per cent of respondents in a survey of preparedness stated that they had *no* intention of preparing for volcanic hazards in the future. Finding that a significant minority of residents are disinclined to anticipate their needing to prepare makes it pertinent to understand this dynamic of citizen engagement in DRR.

The previous chapter introduced how misinterpretation, ignorance and misuse of hazard behaviour data could circumvent people using scientific data to support their anticipating their risk. This chapter expands the range of factors capable of impeding anticipation to include socio-environmental beliefs, social-cognitive biases, emotions and social process. It then discusses the DRR implications of each factor.

3.2 Anticipating risk and preparedness needs

While scientific and risk management professionals can readily anticipate a reality in which significant hazard events will occur in the future, this is not always the case for their citizen counterparts. At the heart of citizens' anticipatory reticence is the fact that large-scale hazard events are complex, threatening, and occur infrequently.

The infrequent nature of large-scale hazard events means that people lack any tangible frame of reference to guide their thinking about their hazardous futures. The absence of concrete experience to guide people's deliberations increases the scope for several social-cognitive, emotional and interpretive biases and processes to affect people's anticipatory thinking. If the reasons underlying people's anticipatory reticence can be articulated, this understanding can inform the development of strategies to motivate people's ability to *anticipate* what they will encounter when disaster strikes and stimulate their engagement in DRR processes.

So, what are these processes and how do they act to undermine people's interest in engaging in DRR processes? Impediments to anticipation can arise from several factors, some of which are surprising.

3.2.1 Embarrassment

For example, McBride, Becker and Johnston (2019) found that anticipatory thinking was curtailed by people feeling embarrassed about being involved in DRR planning activities (in this case, a *drop, cover, and hold* earthquake preparedness drill). Anticipatory reticence can also be influenced by people's interpretation of their socio-environmental relationship.

3.2.2 Socio-environmental perceptions

People (particularly in Western cultures) who conceptualize socio-environmental relationships in anthropocentric terms tend to perceive themselves as independent of nature and believe that people can control nature (Buergelt et al., 2017; Charlesworth & Okereke, 2010; Paton, Buergelt & Campbell, 2015; Woodgate & Redclift, 1998). Such environmental control attributions increase the likelihood of people transferring responsibility for DRR activities from themselves to the risk management and scientific agencies they deem able to exercise control over environmental risk. Transferring responsibility to others reduces the likelihood of people being motivated to anticipate their needs, diminishing the likelihood of their engaging in DRR processes. Additional challenges to anticipation derive from the action of several cognitive biases. The first of these discussed is unrealistic optimism bias.

3.2.3 Unrealistic optimism

Particularly in circumstances in which people lack experience of hazard events, unrealistic optimism drives people to believe that, compared to the average person, they themselves are less likely to suffer future misfortunes (Weinstein, 1980). In a DRR context, this results in people underestimating the likelihood of *their* experiencing a disaster.

While unrealistic optimism does not negate the belief that hazardous events (e.g., earthquakes, hurricanes) will occur in the future, it reduces the likelihood of a person believing that *they* will suffer from such occurrences; they think the consequences of disasters will affect other people but not them (Burton et al., 1993; Spittal et al., 2005). Unrealistic optimism can also manifest in people believing that hazard events (e.g., earthquakes) will happen somewhere else, but not where they live (McClure, 2017).

The net effect of the action of this bias is that people effectively transfer risk from themselves to others in their community (Lindell & Perry, 2000), reducing their perceived need to anticipate their future risk. This, in turn, lessens the likelihood of people believing that risk information is relevant for them, reducing their perceived need to engage in DRR processes.

Strategies to counter the influence of this bias and motivate people's interest in engaging in DRR activities include providing opportunities for people to actively engage in discussing hazard consequences and preparedness with members of comparable communities. Discussions with like-minded others, especially if they can provide accounts of the disaster consequences experienced and the preparedness measures that proved effective, can help disarm the adverse influences of unrealistic optimism (Adams et al., 2017; Lindell & Perry, 2000; Paton & McClure, 2013; Weinstein, 1980).

Other strategies (including those used to counter beliefs that disasters will occur elsewhere) include providing localized hazard information, including local hazard distribution maps and interactive tools, that provide people with more tangible insights into what they will have to contend with in future (McClure, 2017). These strategies are more effective in motivating people's interest in anticipating environmental hazard issues than providing scientific information and advice on risk per se (Chaiken, 1980). Another cognitive bias relevant here is risk compensation.

3.2.4 Risk compensation

People's lack of tangible hazard experience also increases the scope of the risk compensation bias to contribute to anticipatory reticence (Ballantyne et al., 2000; Etkin, 1999; Fischhoff, 1995; Lupton, 1999). The risk compensation bias describes how people's beliefs regarding *their* need to act to enhance their safety are inversely proportional to their perception of the threat they face from environmental hazards. For example, people's awareness of the existence of societal-level mitigation (e.g., building levees, building codes) measures reduces their perception of environmental threat. The latter is accurate to the extent that it reduces local flood risk or increases building capacity to withstand impacts from earthquakes up to a specified intensity – people's interpretation of the effectiveness of these actions may exceed those intended by their risk management counterparts. Risk compensation bias increases the likelihood that people will inappropriately assume that the presence of structural measures (e.g., a levee) eliminates their household exposure to environmental threats (e.g., from flooding). Consequently, it lessens their belief that they need to anticipate future events and undertake any personal or household protective measures. While risk management agencies separate the roles of structural mitigation and household preparedness, and assume that households will continue to take responsibility for their preparedness, risk compensation calls this risk management assumption into question.

For example, in a study of volcanic preparedness, Paton, Smith et al. (2008) found that a volcanic hazard public education program had provided people with information about what scientific and local government agencies were doing to manage volcanic risk that citizens were previously unaware of. While the information received was unrelated to household risk per se, people's receipt of this previously unknown information about societal research and planning, resulted in some 28 per cent of respondents stating their intention to *reduce* their level of preparedness. Obtaining hitherto unknown information about what scientific and local government agencies were doing to manage volcanic risk resulted in people perceiving their environment as being safer than they had previously thought. Even though the information was unrelated to household preparedness (e.g., it covered monitoring, local government training), people's perception of the improvement in their environmental safety resulted in their reducing their need to take responsibility for their household preparedness.

The limited awareness of this relationship within societal hazard planning processes means that this relationship may be missed; risk management authorities can thus be unaware of how risk compensation bias can diminish citizens' interest in anticipating their risk and preparedness needs. To counter the potential for this bias, risk management agencies must take steps to ensure that households and communities remain aware of their shared responsibility for household, neighbourhood and community safety, and continue to adapt and maintain appropriate DRR activities.

Approaches to countering the influence of this bias include providing information in ways that emphasize how societal mitigation (e.g., building codes) and personal/household preparedness (e.g., securing household fittings and fixtures such as the TV) complement (rather than their being substitutable for) one another. It is also important that risk communication makes it explicit that people are responsible for reducing *their* household and neighbourhood risk (Paton, Kerstholt & Skinner, 2017). Social marketing approaches to managing risk compensation (Guion et al., 2007) suggest that providing comprehensive cost and benefit information (e.g., on structural mitigation) and including information that challenges people's overestimation of the role of structural mitigation can reduce the influence of this bias. Challenges to anticipation can also emerge from people's emotional or affective responses to environmental threats.

3.2.5 Denial and fatalism

If people believe they are unable to exercise any control over sources of hazards (e.g., earthquakes), they may cope by denying the seriousness of the risk posed by such environmental hazards, and use denial as a means for managing their hazard related anxiety (Crozier et al., 2006; Gifford et al., 2009). Responding to environmental threats by denying their implications lessens the likelihood that people will engage in anticipatory thinking about events that are the focus of their denial.

This relationship makes the task of communicating about hazard events challenging. Indeed, doing so can be counterproductive; for those with hazard-related anxieties or who are engaging in denial, receiving information about environmental hazards will increase their anxiety and denial, increase their adopting fatalistic approaches to DRR, and diminish the likelihood of their engaging in anticipatory thinking and preparedness (Crozier et al., 2006; McClure, 2017; Paton et al., 2005). Denial, fatalism and anxiety thus pose significant challenges to DRR.

The key to understanding these issues derives from how denial and fatalism are linked to people believing that if it is impossible to prevent events such as earthquakes occurring, then they are powerless to act to protect themselves. While earthquakes themselves cannot be prevented, the same cannot be said for their consequences. Hence, one approach to reducing fatalism and denial involves reframing information by providing people with illustrative examples that help them differentiate causes from consequences and to appreciate that hazard consequences can be reduced or prevented through personal actions (McClure, 2017; Vinnell et al., 2020).

For example, viewing photographs of earthquake damage that shows intact buildings alongside collapsed buildings creates opportunities for people to appreciate that actions people can perform (e.g., regarding building design, construction techniques) can mitigate earthquake damage. Similarly, illustrating how actions such as installing stays on tall furniture or adding flexible hoses to gas appliances can prevent damage to household fixtures and fittings or encouraging involvement in disaster drills can provide tangible cues to how personal actions can prevent losses from hazard consequences (Crozier et al., 2006; Vinnell et al., 2020). An important adjunct to this process involves providing specific information on how each consequence occurs and on how each recommended action (e.g., how the risk posed by the ground shaking created by earthquakes is reduced through structural measures such as strengthening chimneys and securing furniture and fittings) can be implemented (Crozier et al., 2006). The approach advocated by Crozier and colleagues motivates people's realization that they are not helpless, provides them with tangible examples of how control can be exercised over hazard consequences, and helps engage people in DRR activities. A related issue here is hazard-related anxiety.

3.2.6 Anxiety

Anxiety, a future-oriented mood state mobilized by confronting events that are perceived as unpredictable and uncontrollable (e.g., disasters), can mobilize actions to cope with anticipated threats (i.e., it has adaptive capabilities) if people believe they can exercise some control over the events they face (see above). If, however, people feel overwhelmed (e.g., being given too much information, media overemphasis on catastrophic hazard outcomes), their perceived sense of control is diminished and their ensuing feelings of anxiety reduce the likelihood of their engaging in anticipatory DRR thinking (Barlow, 2002; Paton et al., 2005; Kerstholt et al., 2017; Slovic et al., 2002). This relationship is challenging for risk communication; informing people about threatening and unpredictable environmental hazard events can increase dysfunctional anxiety. While anxiety-inducing information can mobilize short-term engagement in hazard preparedness, over the longer term it tends to have the opposite effect (Daniel, 2007; Hastings et al., 2004).

Over the longer term, receiving anxiety- or fear-inducing information makes it more likely that people will ignore information content and may even manage their anxiety by derogating or questioning the expertise of information sources (e.g., emergency management agencies). The net effect is disengagement from the DRR process (Armaş et al., 2017; Kerstholt et al., 2017; McLennan et al., 2014; Mishra & Suar, 2012).

Clues to managing anxiety's influence on DRR participation derives from Witte's (1992) discussion of how it is a combination of high perceived threat and low perceived efficacy in recipients of DRR communication that activates defensive fear-control processes such as denial of risk, information avoidance, and reticence in anticipating future risk and preparedness needs. Consequently, one approach to countering the adverse effects of fear and anxiety on anticipatory thinking involves increasing people's control or perceived control in their DRR deliberations (Crozier et al., 2006; McLennan et al., 2014; Witte, 1992). Including psychological preparedness (that encompasses strategies for anticipating sources of anxiety, identifying distressing thoughts and emotions that may exacerbate anxiety, and developing stress management strategies) strategies can also be effective (Cohen & Abukhalaf, 2021; Morrissey & Reser, 2003). However, the kinds of emotional reactions to risk information outlined above may not arise if people are overconfident in their knowledge or abilities.

3.2.7 Overconfidence

If people are overconfident about their existing knowledge, this can affect their motivation to anticipate their current and future risk and their DRR needs (Slovic et al., 1982). One example of overconfidence emerged in a study of hazard knowledge in New Zealand.

In New Zealand, actions to perform should a volcanic eruption occur are listed in the Yellow Pages telephone directory. In a telephone survey, Ballantyne et al. (2000) asked a sample of 410 Auckland residents, 92 per cent of whom acknowledged their volcanic risk, if they could name these activities. While 41 per cent believed that they could name these actions, when asked to do so, only 6 per cent (of the total sample) could actually do so. This discrepancy (41 per cent who believed they knew them vs. 6 per cent who could actually name them) illustrates how overconfidence can derive from people conflating knowing where relevant information is located with their actual knowledge. Similar findings emerged from work by Ward (2021). Ward found that people can believe that information available from Google is *their* information (rather than information residing on the Internet) and assume it is part of their general knowledge. They thus overestimate *their* hazard knowledge, and this lessens the likelihood of their anticipating their having any future DRR needs.

One approach to countering the deleterious influence of overconfidence involves providing information that challenges the accuracy of people's (overconfident) judgements and provides them with cues for making more realistic judgements (Plous, 1993). Another approach involves creating opportunities for people to take a more critical look at their risk assessments by asking them to list the pros and cons of their beliefs and how these affect their predictions about their future risk (McClure, 2017). This approach counters overconfidence by increasing the likelihood that people will contemplate alterative, more challenging, future events and start to consider their future risk information and preparedness needs accordingly. Another challenge to anticipation can be traced to more fundamental issues regarding the thought processes used to make DRR judgements.

3.2.8 Analytical and affect-driven decision processes

People's risk judgements are influenced by two thought systems; the learned, cognitive, analytic system, which involves slower and more deliberative cognitive processes versus the associative and affect-driven, experiential system, which applies rapid, unconscious affective processes to people's decision-making (Chaiken & Trope, 1999). Scientists and risk management professionals typically apply the analytic processing style to their analysis and interpretation of risk and other DRR data.

In contrast, citizen recipients of DRR information rely more on the associative processing system. This type of processing increases the likelihood of people interpreting information about uncertain and potentially catastrophic environmental hazard events in terms of affective responses, such as fear or anxiety (Chaiken & Trope, 1999). Information that stimulates emotional responses such as fear, feeling overwhelmed, or feeling powerlessness kindles the development of negative beliefs about hazard events and their associated DRR processes, reducing people's interest in deliberatively anticipating their future DRR needs and circumstances. There are other ways in which the emotional undermining of anticipation can occur.

3.2.9 Emotion and anticipation

Other ways in which people's engagement with DRR processes can culminate in some experiencing negative emotions such as fear and anxiety include being presented with detailed lists of preparedness items. In addition to overwhelming people, the inclusion of information about high-cost actions (e.g., need for structural retrofitting homes in earthquake-prone areas) can trigger feelings of dread and anxiety (Crozier et al., 2006; McClure et al., 2001). These authors describe how exposure to information with high-cost (e.g., financial, time, opportunity costs etc.) implications elicits strong negative emotional responses that impede anticipatory thinking by making it more likely that people will chose to ignore risk data, including information that describes the potential benefits that can accrue from structural preparedness (e.g., increasing the value of their home). A related issue derives from differences in when costs are incurred and when benefits could be realized.

3.3 Temporal influences on cost and benefit judgements

People's anticipation of preparedness needs is affected by the fact that the costs of preparedness are immediate but any benefits will only be realized at some indeterminate time in the future (i.e., when disaster strikes). In this context, people's perception of long-term versus immediate or short-term tangible costs makes them less likely to contemplate future benefits, and this affects how they engage with DRR processes.

For example, Paton, Kelly et al. (2006) found, when researching wildfire preparedness, that while some people routinely prepared at the start of each fire season, others would do so only when an imminent threat (e.g., fire could threaten their property) was evident. The latter decision derived from people's cost–benefit judgements; the benefits of preparing were perceived to outweigh the costs when an actual threat was present. Prior to a fire being present in their environment, the immediate costs would outweigh benefits that were not perceived as likely to occur until some indeterminate time in the future. Decision-making of this kind is also affected by how future benefits are perceived as more abstract and so less likely to motivate people's DRR engagement (McClure, 2017).

The reality is that this type of cost–benefit judgement would leave residents with insufficient time to prepare (particularly regarding, for instance, creating a defensible space, securing eaves and openings to prevent ember inundation, etc.). This judgemental process was also influenced by people misinterpreting (see Chapter 2) warning and response times (Prior & Paton, 2008).

Prior and Paton (2008) found that people who (inappropriately) believed that warnings would provide a day or more lead time were more inclined to say that they would not prepare until they received a warning. However, given a reality in which fire warnings precede response times measured in hours rather than days, making incorrect assumptions about the relationship between receipt of a warning and available response times increases people's risk and results in a greater likelihood of people delaying acting until it is too late.

Approaches to countering these impediments include highlighting in public education and risk communication information that warnings should not be used as triggers for preparing, and highlighting that response times of hours make pre-event preparedness a DRR imperative (Prior & Paton, 2008). Similar approaches have been advocated for volcanic hazards (Gregg et al., 2004).

The above discussion identified how anticipation can be undermined in several ways, and this circumstance requires adopting strategies to overcome these constraints. The goal here is to create the circumstances in which people can begin to anticipate their hazardous circumstances, the risks they are exposed to, and their preparedness needs. It is also important to note that getting people to anticipate their risk does not, in itself, motivate or mobilize actual preparedness. To do so, attention must turn to understanding how to facilitate people's ability to prepare. This issue is tackled in the next chapter.

4 Preparedness

> Tell me, and I will forget,
> Show me, and I may remember,
> Involve me, and I will understand.
> Confucius (551–479 BCE)

4.1 Preparing

The preceding chapter identified how a substantial minority of people may fail to engage in DRR processes and discussed strategies available to counter these constraints. Countering these constraints lays a foundation for enabling people to convert their anticipatory thinking into actual preparedness, a key component of a comprehensive DRR strategy (UNDRR, 2015, 2017).

The UNDRR (2016) defines preparedness as the *knowledge and capacities developed* by governments, response and recovery organizations, communities and individuals to effectively anticipate, respond to and recover from the impacts of *likely, imminent or current* disasters. This chapter introduces how the *knowledge* and *capacities* people require to effectively respond to *likely events* are developed.

Two lines from the Confucian epigraph that opens this chapter resonate with contemporary perspectives on preparedness. The first, "tell me, and I will forget," suggests that just making information available to people may not mobilize preparedness. The second, "involve me, and I will understand," introduces how collaborative and cooperative strategies that emphasize shared responsibility are more likely to result in action. Before discussing the insights offered by these contrasting perspectives, this chapter first introduces what being prepared means.

4.2 What does being prepared mean?

There is no clear definition what comprehensive preparedness is. This is not surprising, at least in terms of content. Some aspects of preparedness are the same, irrespective of the source of a hazard. For instance, people need to be self-reliant (e.g., stored food and water) during initial periods of disruption from any event. However, when it comes to structural preparedness, the source of a hazard has implications for preparedness. For example, structural preparedness for earthquakes calls for measures (e.g., retrofitting vulnerable features such as chimneys, selecting building materials, securing internal fittings and fixtures) that differ from those needed to safeguard the home from volcanic hazards (e.g., steeper roof profiles, removal of gutters to limit loss from ashfall, covering downpipes to prevent blockage from volcanic ash). For this chapter, the focus is less on specific items of content and more on introducing preparedness in terms of the functions it fulfils.

Comprehensive preparedness comprises *structural* (e.g., seismic retrofitting, creating a defensible space), *survival* (e.g., storing food and water), *planning* (e.g., household planning, attending community meetings), *psychological* (e.g., enhance self and family coping), *community/ capacity-building* (e.g., work with neighbours to confront local issues), *livelihood* (e.g., meet employment needs), and *community-agency* (e.g., collaborate with response agencies, businesses, NGOs) functional categories (Lindell et al., 2009; Paton, Anderson, Becker & Peterson, 2015; Paton & McClure, 2017). Other researchers have argued for *evacuation* preparedness to be included in this list (Onuma et al., 2017). These functional categories are summarized in Table 4.1. An important issue is that, to optimize safety and resilience, the comprehensive adoption of each functional category is required.

4.3 Preparedness and resilience

The functional preparedness categories listed in Table 4.1 have significant implications for household, community and societal resilience. For example, households that have adopted relevant structural (e.g., secured home to its foundation (earthquake), installed roof clips to mitigate hurricane impacts, elevated homes in flood-prone area) and survival (e.g.,

Table 4.1 Functional preparedness categories

Functional Preparedness Category	Personal, Household and Community Activities (illustrative examples)
Structural	Constructing a defensible space around the home and covering ventilation grills and eves to prevent embers entering the home (wildfire), securing a home to its foundations and internal fixtures and fittings to walls (earthquakes), sealing off gutters and downpipes to prevent their blockage by volcanic ash, elevating the ground floor to minimize flood inundation ...
Psychological	Identifying potential stressors in self, family and others and rehearsing coping strategies for self, children and family, anticipating stress management issues over time (e.g., dealing with aftershocks, evacuation, relocation), developing social support strategies ...
Survival/Direct Action	Storing water, food (for each household member), battery radio and spare batteries, essential medicines, etc. to enable response to loss of utilities and absence of support from authorities ...
Planning (Household, Family and Personal)	Acquiring hazard and preparedness knowledge, using these to develop family capabilities and response plans to ensure all know how to respond, where to go, how to make contact if family members separated, contingency planning (e.g., disaster occurs in winter or summer, parents at work, children at school) ...
Community-/ Capacity-building	Actively participating in hazard-related community meetings (neighbourhood, suburb, etc.), contributing to neighbourhood planning, compiling inventories of relevant neighbourhood experiences, skills and resources and their local response and recovery application, developing an inventory of vulnerable people in the neighbourhood and planning how they can be assisted and supported. Identifying local community leaders ...

Functional Preparedness Category	Personal, Household and Community Activities (illustrative examples)
Livelihood	Anticipating short- and long-term disruptions to employment, involvement in business continuity planning, planning to manage disruptions, working from home or planning alternative working arrangements, ensure these complement household preparedness and plans …
Community Agency	Attend/initiate meetings with local government departments, businesses, NGOs, and agencies with recovery responsibilities, develop liaison mechanisms where possible to plan how government agencies can support local planning and needs, inviting risk management and scientific personnel to discuss local issues, provide input into local planning …

Source: Adapted from Paton, Kerstholt & Skinner, 2017.

food and water available for each family member) preparedness measures are more likely to be able to remain in situ and be more self-reliant when disaster strikes.

The adoption of structural measures (e.g., secure home to foundations) reduces the likelihood of injury, incapacitation and death to family members and increases the likelihood that people will have an undamaged, or at least habitable, home to remain in or return to. Family self-reliance is enhanced by having access to survival resources (e.g., food, water, torch, radio, medicines, alternative cooking sources) that enable their being able to respond with no or less dependence on societal assistance. Psychological preparedness can facilitate personal and family coping and enable their capability to provide social support to family members, neighbours, and members of their community or social networks. Finally, the pre-event development of community/neighbourhood plans and skill and resource inventories can expedite the collaborative provision of mutual assistance and enable local recovery independently of societal assistance.

If people can remain in situ (by virtue of their being structurally, personally and socially prepared), they increase their availability to enable their and their community's resilience by being available to participate in local recovery activities independently of societal assistance and by being able to contribute (e.g., as employees, as consumers, as volunteers) to regenerating local social and economic recovery (e.g., to support economic activities as employees and consumers). Furthermore, the more people can function effectively without recourse to formal societal assistance, the more the limited supply of formal recovery resources can be dedicated to, for example, restoring lifelines, repairing infrastructure and meeting the needs of more vulnerable members of society. Comprehensive preparedness is thus crucial to developing and maintaining personal, household, social, societal and economic resilience in disaster-affected areas.

In contrast, households who fail to adopt relevant structural and survival measures face a greater likelihood of evacuation and becoming dependent on formal and NGO agencies to meet their survival needs. Families who must evacuate face greater levels of stress and lose access to social support from neighbours, friends and work colleagues.

Moreover, the greater the proportion of ill-prepared people within a jurisdiction, the more resources formal and NGO agencies will have to direct to meeting the needs of people who could otherwise have supported themselves. The latter circumstance dilutes the resources available to support those more vulnerable members of a community whose circumstances limit their ability to prepare and who should thus be prioritized in response settings. Being ill-prepared thus reduces personal and social resilience and diverts response resources from tasks that would otherwise support a more resilient response.

Despite the advantages that can accrue from doing so, being comprehensively prepared is the exception rather than the rule. This makes it important to understand why levels of preparedness remain low and to inquire how it can be developed. This discussion starts with asking whether the answer lies with people's lack of access to relevant information and thus whether giving them more information is the key to increasing preparedness.

4.4 Preparedness information

There is a long-standing assumption that people's preparatory reticence derives from a lack of relevant information (e.g., via public education, social media). This assumption reflects the influence of the so-called *knowledge deficit model* (Arneson et al., 2017; Simis et al., 2016). The knowledge deficit model assumes that redressing people's preparatory reticence entails providing them with more information. Challenges to the validity of this view are, however, not new.

The quote from Confucius, penned some 2,500 years ago, describes how if you "tell me ... I will forget." The inadequacies of relying on providing information to change behaviour have thus been known for millennia. While it is undeniable that information about risk and preparedness is important (Lindell & Perry, 2000, 2012), it represents only one input into the preparedness process. Others relate to, for example, how people construe their risk and interpret their needs, with these processes introducing a need to understand how people first interpret their circumstances and then determine what information is relevant for them. The inclusion of interpretive processes in this regard introduces the possibility that people may interpret some information made generally available by risk management sources as irrelevant to their needs; it is how people impose meaning on information that is important.

However, even if information is deemed relevant, if people determine that they lack the skills or resources to act on it, information will be ignored. A more searching analysis of preparedness processes is thus required.

4.5 Understanding and facilitating preparedness

People differ regarding the quality (i.e., number of functional categories adopted) and quantity (i.e., number of items from each functional category) of their preparedness, and these differences occur even when people acknowledge their risk (Harries, 2008; Johnson & Nakayachi, 2017; Lindell et al., 2009; Marti et al., 2018; Onuma et al., 2017; Paton et al., 2005; Wachinger et al., 2012). The starting point for exploring preparedness derives from people's preference for focusing their preparedness efforts on survival (e.g., storing food and water) preparedness and are generally less likely to consider its structural and community relationship counterparts (Paton & McClure, 2013).

While survival preparedness is undeniably important, if an absence of adequate structural preparedness results in death or injury, survival resources are rendered useless. Understanding this predisposition towards survival preparedness is thus important, as is determining how to encourage expanding people's preparedness inventories to include, for example, structural and community preparedness and to enable the sustained adoption of more comprehensive levels of preparedness.

4.5.1 Favouring low-cost preparedness

This section introduces why people favour adopting low-cost, easily implemented (e.g., survival) items. Understanding why can identify what needs to be done to motivate people to shift from this position and progressively adopt more comprehensive, but higher-cost, preparedness (e.g., structural, community relationship preparedness) measures. The first question concerns what lies at the roots of favouring low-cost items.

This discussion introduces the *single action bias* as an explanatory mechanism proposed to account for people preferencing the adoption of low-cost, easily adopted measures (e.g., survival preparedness) and for neglecting to consider adopting higher-cost (e.g., structural, community relationship preparedness) actions (McClure et al., 2009). This bias describes how, when presented with comprehensive lists containing, for example, survival, structural, and community relationship items (that differ considerably in financial, time and social costs involved in adoption), people focus on a single, low-cost option and discount the value of others, and especially the more costly alternatives (McClure et al., 2009). This bias lessens the likelihood of people considering any other items (Finucane et al., 2000; McClure et al., 2009; Slovic et al., 2002).

An important adjunct to understanding people's preferencing low-cost survival items derives from how attending to low-cost items elicits positive emotions (see Chapter 3). The experience of these positive emotions sustains people's tendency to continue favouring low-cost options (e.g., Finucane et al., 2000; Slovic et al., 2002). Furthermore, these authors describe how presenting information on high-cost items has the opposite effect; they elicit negative emotions (see Chapter 3). The negative emotions generated when faced with high-cost options lessen the likelihood of people considering adopting them or even contemplating the benefits that may accrue from their adoption (e.g., how structural measures enhance the value of the family home).

McClure et al. (2009) suggest that overcoming these low-cost biases starts with breaking down complex, multi-category lists into several lists containing two or three items from each functional category (Table 4.1) and progressively introducing these smaller lists to people over time. The process starts with presenting multi-purpose, low-cost items. Once adopted, the process moves to progressively presenting more complex and costly (e.g., structural changes, engaging an engineer to check the structural integrity of a house) items. This strategy has been endorsed by others (e.g., Lindell & Perry, 2000).

An important adjunct to this process is the contents of each list being accompanied by specific information about how each measure can mitigate adverse hazard consequences and including information that describes the additional benefits that can accrue from preparedness (Guion et al., 2007; McClure et al., 2009). For example, when communicating about structural preparedness, discussions can progress beyond providing information on how it mitigates the impact of, for example, ground shaking or ember attack, to include advising people how structural measures can enhance family safety and improve the value of the home (DiPasquale & Glaeser, 1999; Vinnell et al., 2019). The latter course of action may help ameliorate the mobilization of negative emotions but this is an area requiring additional work. Other explanations for favouring low-cost options have been proposed. One derives from several low-cost, survival items being present in households independently of their preparedness roles.

For example, people can have a stock of food in the house and a portable radio they use every day. However, people can conflate the presence of such everyday items with their survival functions. If this occurs, people's interest in advancing their level of disaster preparedness is diminished.

For example, in a preparedness survey, Paton (2008) asked people "do you have 3 days' supply of tinned food". Some 67 per cent of respondents answered in the affirmative. However, when a subsequent question asked people if they had *changed their shopping habits to gradually increase their emergency food supplies*, a different outcome emerged. For the second question, only 16 per cent of respondents described themselves as building their disaster-specific preparedness food supply.

The discrepancy (67 per cent vs. 16 per cent) between these results demonstrates how people could conflate food availability derived from their shopping habits with food being available and set aside in their disaster survival kit. While people who conflate, for instance, acquiring food via their regular shopping habits with having an emergency food supply, could have adequate food resource available if a disaster occurred, this outcome could be attributed more to luck (e.g., if a disaster occurred immediately after a shopping trip) than good preparedness decisions. Overestimating preparedness can arise for other reasons.

Knowledge of people's preparedness is frequently obtained using self-report questionnaire data. People are more likely to describe their survival preparedness in these studies. However, subsequent audits of people's questionnaire responses revealed that people can overestimate their actual level of preparedness.

Lopes (1992) found that people's recall of their level of preparedness derived from their remembering what they had done in the past. For instance, people would remember compiling a survival kit in the past but forgot periodically removing items to meet routine needs (e.g., getting batteries, tinned food). However, despite not replenishing the removed items, people completed the questionnaire based on their recall of their earlier actions rather than their current circumstances. They thus overestimated their preparedness. This calls for caution when relying on self-report preparedness data. If people overestimate their level of preparedness, they are less likely to perceive a need to develop it (Grothmann & Reusswig, 2006). These examples highlight the importance of outreach strategies reminding people to check and maintain their disaster-specific survival resources (Becker et al., 2013).

People's overconfidence regarding the adequacy of survival preparedness can arise following experience of relatively low-intensity events. If low-level hazard experience does not challenge people's existing level of preparedness, they extrapolate from their being unaffected by the low-intensity experience a capacity to respond to more intense events without any additional preparedness or need to attend to risk information (Johnston et al., 1999).

Rectifying this problem involves specifically drawing attention to the range of intensities that could occur, the difficulties inherent in predicting these in advance, and informing people of the specific preparedness needs associated with different levels of hazard intensity (McClure et al., 2009). The effectiveness of this approach can be heightened using interactive techniques, such as hazard intensity and distribution maps, to facilitate discussion of the local risk management and preparedness associated with high-intensity events. These techniques can be used to illustrate how hazard intensity, distribution and duration (see Chapter 2) interact and to stimulate discussion of the implications of these data for local and household planning and preparedness.

The preceding discussion introduced strategies that can challenge people's preference for low-cost survival preparedness and so lay the foundations for advancing towards more comprehensive levels of preparedness. To sustain this kind of preparedness momentum, other strategies must be brought into play. In the next section, these strategies are introduced using preparedness theories.

4.6 Preparedness theories

Several theories have been developed to account for differences in levels of comprehensive preparedness. While preparedness theories adopt different routes to arriving at this destination, the variables adopted describe the interdependent roles environmental perceptions (e.g., risk perception), social interpretive processes (e.g., social capital, empowerment), and personal and collective competencies (e.g., self-efficacy, collective efficacy) and information access play in people's preparedness judgements. The dependent variables used when testing preparedness theories comprise, to a greater or lesser extent, items representative of the functional categories described in Table 4.1. For some functional categories, item content and wording are adapted to suit the hazard consequences being investigated. This is most pronounced for structural preparedness.

For instance, structural preparedness for wildfires includes creating defensible space and preventing embers entering a home while its earthquake counterpart includes structural reinforcement and securing fixtures and fittings to walls. The accommodation of these differences in theory testing has ensured that the theories discussed here have demonstrated their predictive utility against multiple hazards. The latter capability is reinforced by an important characteristic of the independent variables used to populate these theories – their ability to be conceptualized as adaptive capacities.

Variables that tap into how people's environmental threat beliefs (e.g., risk perception, perceived severity of the threat) relate more specifically to natural hazard phenomena. However, others have a more ubiquitous applicability; they describe capacities that facilitate people's ability to adapt to any challenging circumstances. For example, self- and collective efficacy are capacities that enhance people's ability to plan how to confront any challenges encountered and the higher the scores, the greater is people's persistence in dealing with the challenges they encounter. Constructs such as community participation and bonding and bridging social capital enable people to develop their collective understanding of a problem and to collaborate to determine how best to deal with it (irrespective of whether people are seeking to get a new park in a neighbourhood, mounting local opposition to a new road, or planning for disaster).

These examples describe capacities that enable people to adapt to any circumstances. Hence, they can be described as adaptive capacities. This situational flexibility, and the inclusion of such variables in preparedness theories, contributes to their being able to meet the all-hazards requirement established by the UNDRR (2015). The discussion of these theories starts with the Health Belief Model.

The Health Belief Model (HBM) proposes that if people's assessment of their *perceived susceptibility to threat* combined with the perceived *severity of the threat* is sufficiently strong, their decision to prepare is mediated by their assessment of the *personal costs and benefits* of preparing. The HBM has been used to predict earthquake and flood preparedness (Dooley et al., 1992; Ejeta et al., 2016).

Another theory, Protection Motivation Theory (PMT), has been applied to studies of wildfire and flood preparedness (Grothmann & Reusswig, 2006; Martin et al., 2007; McLennan et al., 2014). The PMT argues that people's motivation to prepare derives from their first personalizing and accepting their risk from environmental hazards. The relationship between this risk acceptance and preparedness is mediated by people's assessment of both the probable efficacy of preparedness actions, and people's assessment of their personal ability to implement them (e.g., their assessment of their *self-efficacy* and *coping appraisal*).

Another effective theory, the Person-relative-to-Event (PrE) theory (Duval & Mulilis, 1999; Mulilis & Duval, 1995), describes how *perceived resource availability* mediates the relationship between people's assessment of personal competencies (e.g., *coping appraisal, self-efficacy*) and their expectancies about the hazardous events they could experience (e.g., beliefs regarding severity) and preparing. The PrE theory has been applied to accounting for differences in levels of earthquake (Duval & Mulilis, 1999; Hu et al., 2022; Mulilis & Duval, 1995) and tornado (Mulilis et al., 2000; Mulilis et al., 2003) preparedness.

Other theories have expanded the range of variables canvassed to include attitudinal and normative factors and, in so doing, introduce how social context variables influence preparedness outcomes. Doing so reiterates the wisdom in the "involve me, and I will understand" line from Confucius (opening epigraph).

Confucius' identification of the relationship between social involvement and understanding is one echoed in the several theories that identify how collaborative, cooperative and social constructionist processes influence preparedness. The inclusion of strategies based on involving people is important.

This section now turns to discussing how theories that encompass social context and social process variables help understand how people interpret, impose meaning on, and make preparedness decisions about, the hazardous circumstances they could encounter. The importance of including social context and social process variables is reinforced by finding that people's preparedness decisions are influenced more by others in their social networks than by information from scientific and risk management expert sources per se (Adams et al., 2017; Aldrich & Meyer, 2015; Andreason, 2007; McLennan et al., 2014; Mileti & Darlington, 1997; Paton, 2008, 2013; Paton et al., 2005; Terpstra & Lindell, 2013).

One theory that includes a social component, the Theory of Planned Behaviour (TPB), describes the interdependent contributions that people's attitudes to preparedness, their preparedness norms (e.g., whether they think significant social others are favourably disposed to preparing), and their appraisal of whether they possess the level of behavioural control required to prepare accounts for differences in levels of people's preparedness. The TPB has demonstrated a capacity to predict preparedness for flood, wildfire and earthquake preparedness (McIvor & Paton, 2007; McLennan et al., 2014; Slotter et al., 2020; Vinnell et al., 2019). Another approach to including social processes is evident in Critical Awareness (CA) theory.

Critical Awareness (Dalton et al., 2001) is a measure of the extent to which people think about and discuss with significant others issues of personal and collective significance. This construct was incorporated into a theory that proposed that Critical Awareness (i.e., frequency of discussing and thinking about hazard and preparedness issues with other social network members) predicts preparedness. The relationships between Critical Awareness, risk perception and hazard-specific anxiety and preparedness was mediated by resource (e.g., time, skill) availability, self-efficacy, personal responsibility and problem-focused coping, and moderated by trust in sources of DRR information. The CA theory has been used to predict earthquake and wildfire preparedness (Paton et al., 2005; Paton, Kelly et al., 2006). Other contributions to DRR that emphasize social influences include community-based social marketing.

Community-based social marketing combines marketing and psychological behaviour change constructs (e.g., to counter fatalism, knowledge deficits) to develop risk communication plans (McKenzie-Mohr, 2000). A significant feature of community-based social marketing strategies is its emphasis on accommodating social diversity in different community groups and sub-groups (Andreason, 2007; Faulkner & Ball, 2007; Guion et al., 2007; McKenzie-Mohr, 2000; McKenzie-Mohr & Smith, 1999).

Community-based social marketing strategies draw attention to how motivating shared responsibility is more likely to occur when risk communication strategies include explaining, illustrating and applying hazard knowledge in ways that focus on accommodating social diversity and less on communicating about technical aspects of risk management (Guion et al., 2007). The latter thus represents a strategy capable of enabling people to personalize the information used to make their DRR choices. This is not the only theory that seeks to integrate DRR information and decision-making processes. Another is the Protective Action Decision Model (PADM) (Lindell & Perry, 2012).

The PADM describes preparedness as a process that combines two elements. The first concerns people's perceptions of environmental threat and its implications for their finding actions available and capable of protecting them from the hazard consequences they identified. This is accompanied with a complementary information search and evaluation (assessing information needs, identifying sources, and assessing when information is needed) process (Lindell & Perry, 2012; Terpstra & Lindell, 2013). The *protection motivation* beliefs that derive from people combining the outcomes of these processes influence the emergence of people's *protective action implementation* and their *preparedness* outcomes.

An important facet of the PADM is its highlighting the importance of understanding and accommodating how relationships between social network members and between them and risk management agencies influence preparedness. Other preparedness theories have expanded the range of social context variables included. Two of these are Community Engagement Theory and social capital.

The Community Engagement Theory (CET; see Paton, 2008, 2013) describes how preparedness starts with people's hazard cognitions regarding whether preparedness actions will or will not mitigate their risk or protect them from hazard consequences. These are encapsulated in people's *outcome expectancy beliefs* (Paton, 2008). If people's environmental cognitions focus on the uncontrollable causes (e.g., earthquakes) of events, negative outcome expectancies arise and diminish the likelihood of their preparing. If, however, people focus on potentially controllable consequences (e.g., effects of ground shaking on houses), the ensuing positive outcome expectancies motivate people's interest in finding out what they might do to manage their risk. In the uncertain context in which preparedness occurs, this includes turning to others to identify potential preparedness options.

While it might be presumed that expert sources would be people's first port of call in this context, expert sources are often relegated to playing secondary roles. Instead, people often prefer to turn first to the members of the social networks with whom they regularly engage with, identify with, and have some affinity with (e.g., in local, social, work groups) to facilitate how they socially construct their risk beliefs and ideas about how to manage their risk (Lion et al., 2002; Paton, McClure & Buergelt, 2006; Paton, 2008; Rippl, 2002).

In the CET, the *community participation* variable captures the level of people's engagement with the like-minded others they believe can help them articulate their risk beliefs and help them decide what they could do to manage their risk. The next stage involves planning what to do and how to act, with people's capacity to do so being informed by their experience in identifying and resolving issues of collective local interest (e.g., neighbourhood safety planning, developing community campaigns). This is captured by the *collective efficacy* variable (Paton, 2008). However, the CET acknowledges that people do remain reliant on external professional and scientific sources of expertise.

Given the uncertainty implicit in the hazardscapes they inhabit, and the general lack of hazard experience this creates, people turn to expert sources to acquire the information, resources and guidance they believe they need to fill gaps in community knowledge. However, the nature of this relationship is complicated by it being influenced less by the information possessed by expert sources per se, and more by people's perceptions of the *quality* of their relationships with risk management and scientific agency sources. These issues are discussed in some detail here as they have additional relevance for understanding how social settings and their characteristics influence preparedness.

Given their expertise and access to relevant information, it could be assumed that people would willingly accept and use the information provided by scientific and risk management sources. This is not, however, always so.

People's decisions about using or not using information from risk management, government and scientific agencies is informed by how their interpretations of their past experiences with a given source influence how much they trust an agency source (Lion et al., 2002; Paton, 2008; Siegrist & Cvetkovich, 2000). Hence, people's direct (e.g., perceptions of whether past experiences marginalized, obstructed, disempowered or empowered them in some way) and indirect (e.g., media coverage of inadequate agency responses in the past) experience of scientific and risk management agencies determines whether agency information will be used. Furthermore, these perceptions are developed independently of the quality of the information obtainable per se (Basolo et al., 2009; Johnson & Nakayachi, 2017; Paton, 2008; Siegrist & Cvetkovich, 2000). A key construct here is empowerment.

Empowerment is an intentional, enduring community-centred process that encompasses mutual respect, critical reflection, caring, and group participation. The presence of these characteristics in social settings enables people to gain meaningful access to and control over the resources they require to sustain their well-being and functionality through democratic participation in the life of their community (Zimmerman, 1992). The importance of empowerment in the preparedness process derives from its influence on determining how much people trust a source of information.

If people believe that an agency (source of information) has consistently acted to *empower* them and the communities they are members of over time, they are more likely to *trust* the source and use the information they provide to inform their preparedness choices (Paton, 2008). In the CET, empowerment and trust variables mediate the relationship between outcome expectancy and community interpretive processes to predict preparedness. The CET has been applied to predicting preparedness for earthquake, volcanic, tsunami, wildfire and flood preparedness (Frandsen et al., 2012; Paton, Buergelt & Prior, 2008; Paton, Smith et al., 2008; Paton, 2013; Ranjbar et al., 2021). Another body of work emphasizing the role of social context factors is the social capital paradigm.

Social capital research reinforces the valuable roles that trust, social norms (particularly regarding reciprocity), and social network characteristics play in facilitating preparedness (Aldrich & Meyer, 2015; Nakagawa & Shaw, 2004). Reininger et al. (2013) added a need to include perceived fairness as a variable in this context. The latter is important in relation to SFDRR Principles that call for DRR decision-making to encompass local risk profiles, facilitate inclusivity, and accommodate and account for social diversity (see Chapter 1).

Essentially, social capital comprises bonding (e.g., bonds amongst family members, members of ethnic groups), bridging (relationships involving friends, members of one's social network, etc.) and linking (e.g., relationships between citizens and government agencies) components. Family contributions to bonding social capital encompass the degree to which members hold similar views about preparedness (Cottrell, 2006). Cottrell discussed how family conflict regarding the need for, or benefit of, preparing can reduce the likelihood of overall family support for preparedness in the home; family dynamics can thus facilitate or constrain preparedness.

Bonding social capital has been linked to enhancing warning effectiveness and hazard preparedness (Hawkins & Maurer, 2010; Sadeka et al., 2015). Sadeka et al. raised an interesting issue regarding the reciprocal relationship between preparedness and social capital. They pointed out how adopting hazard preparedness measures can stimulate the development of social capital, increasing its availability to support future capacity-building (e.g., BBB) programs. Other social and situational characteristics have been implicated in understanding preparedness dynamics.

Variables such as social cohesion, place attachment and identity, people's sense of connectedness to people and place, and sense of community (De Dominicis et al., 2015; Frandsen et al., 2012; Paton, Buergelt & Prior, 2008) have demonstrated their ability to account for differences in levels of preparedness. Other factors used to predict preparedness include social responsibility (Guion et al., 2007) and people's sense of bondedness to their neighbourhood (Heller et al., 2005). The supporting and motivating aspects of social relationships should not, however, be regarded as a fait accompli in preparedness research.

An important social dynamic that must be considered here is the role socio-cultural diversity plays in DRR planning (Elliott et al., 2010; Guion et al., 2007). A failure to accommodate socio-cultural diversity, according to these authors, increases the risk of inter-group conflict arising and undermining DRR processes. Nor is it only inter-group dynamics that are relevant here – intra-group dynamics also deserve consideration.

The theory of Group Faultlines (Lau & Murnighan, 1998, 2005) describes how intra-group processes can occur in social groups (e.g., family, neighbourhood) that have previously enjoyed neutral or cohesive relationships experiencing dysfunctional conflict when an external stress changes the relative importance different group members to attitudes that relate to DRR processes. For instance, research into wildfire preparedness identified how calls from risk management agencies for community-wide property vegetation clearing activated social fragmentation between neighbours who previously enjoyed neutral or positive relations (Paton & Buergelt, 2012).

Prior to this call, all residents held attitudes towards issues such as environmental protection and household safety. Under normal circumstances, these beliefs did not affect the quality of relationships between neighbours. This changed, however, when the local government called for neighbourhood-wide vegetation clearing as part of a fire management strategy. This call constituted a social stressor that changed the relative salience neighbours attributed to their 'safety' and 'environmental protection' beliefs, with this differential attitude mobilization affecting the quality of relationships between neighbours.

Consequently, neighbours who attributed greater salience to their environmental protection (e.g., the need to retain trees and vegetation on their property) beliefs than to their household safety (this occurred prior to the fire season so there was no threat of fire at this time) found themselves in conflict with neighbours for whom family safety was the more salient belief. This conflict had enduring implications; it reduced community cohesion (an important predictor of collective wildfire mitigation and preparedness), created social fragmentation, and lessened support for collective wildfire preparedness over the longer term.

Fault-lines are also more likely to be elicited by strategies that ignore socio-cultural diversity in DRR planning (Elliott et al., 2010; Gregg et al., 2008; Guion et al., 2007) making this area one that merits additional attention in future research, particularly in light of SFDRR Principles calling for DRR processes to be inclusive and to accommodate diversity within local risk assessments and to safeguard against exclusion, marginalization and discrimination.

The kinds of attitude and attitude salience issues that Faultline Theory taps into are not currently considered in either risk management or community assessment processes. Consequently, risk management agencies would be unaware of this social dynamic and its implications for undermining the DRR outcomes they sought to encourage.

A significant issue here is the challenges posed by trying to identify potential fault-lines. Under normal circumstances, there may be few clues regarding diverse underlying attitudes and beliefs in a community and no clear guidelines for identifying what might trigger their differential mobilization. Responding to the latter, Paton and Buergelt (2019) proposed that community assessment and planning using local scenario planning strategies could be helpful in this context.

The kinds of attitudinal and socio-cultural diversity issues discussed here introduce a need to consider how social justice processes can affect several DRR processes, including, for example, information needs and provision, environmental perceptions, social interpretive processes, and the quality of relationships within neighbourhoods and social networks. These issues raise issues about the role of governance in preparedness.

4.7 Preparedness and governance

The discussion of preparedness theories reveals that it aligns well with what SFDRR calls for DRR processes to build on cooperative and collaborative principles. However, given the complex web of diverse socio-cultural-environmental influences that must be accommodated in preparedness planning and intervention, it is pertinent to raise questions regarding the role of governance here. Adaptive governance principles can provide insights into how the complex multi-level, multi-stakeholder socio-cultural-environmental interrelationships introduced above can be accommodated in DRR preparedness planning and intervention (Djalante et al., 2011; Munene et al., 2018).

The relevance of adaptive governance in this context is reinforced by Djalante et al.'s (2011) identification of how its emphasis on building on multi-layered institutional relationships increases its suitability for developing locally relevant and integrated stakeholder actions that integrate governance systems across the local, regional and national levels of government that become involved in developing preparedness policies and its local implementation. This resonates with the inclusion of factors such as empowerment and trust in theories such as CET and social capital. This provides opportunities for devolving governance to include local processes better suited to capitalizing on how empowerment and trust can be used to accommodate local diversity and social dynamic issues (see above) when considering how governance policies and practices can facilitate effective preparedness.

Djalante et al.'s (2011) additional comments regarding how adaptive governance offers a framework in which complementary roles for diverse stakeholders are facilitated and sustained through developing self-organization capability and networks adds to its utility (cf., community empowerment and accommodating the social dynamics introduced in social capital, CET and PADM – see above). Finally, the potential of adaptive governance to accommodate not only the interdependencies between people, environment and hazards that influence preparedness but also how they change over time adds to its applicability in this context (Djalante et al., 2011; Munene et al., 2018). However, if these suggestions are taken on board the governance development process, another adaptive challenge for governance stems from the range of preparedness theories available to support DRR.

The preparedness theories introduced above combine environmental interpretive (e.g., *risk perception*, *outcome expectancy*), personal competence (e.g., *coping*, *self-efficacy*), affective (e.g., *anxiety*) and social, cultural (see Chapter 5) and societal relationship (e.g., *bridging social capital*, *social norms*, *collective efficacy*, *empowerment*, *trust*) in various ways to account for differences in levels of preparedness. Some (e.g., PMT) place more emphasis on environment–person intervention. Others (e.g., CET), focus on environment–social network/community intervention. This creates governance and implementation challenges regarding, for example, the relative merits of theories that favour person–environment level versus community–environment level versus agency–community interventions, or whether all are important. There is, however, an alternative approach.

Given that all the theories introduced above have been empirically validated, rather than seeing them as standalone options in which decisions would be required regarding the mix of theories required in a given setting, an alternative is to integrate the variables from preparedness theories, and those introduced in Chapter 3, to create a matrix approach to modelling the preparedness process. From a governance perspective, this affords opportunities to support comprehensive assessment, design, planning and intervention strategies through selecting options best suited to local needs and circumstances rather than by advocating the use of a specific theory or theories.

Support for this option comes from Adhikari et al.'s (2018) selecting and integrating variables drawn from PMT and CET theories to predict preparedness intentions in populations affected by the 2015 Nepal earthquake. The rationale for Adhikari et al.'s work derived from appreciating that variables such as risk perception and coping maintain inconsistent relationships with pre-event preparedness (Wachinger et al., 2012), with this resulting in these variables being absent from the CET. However, in the recovery settings in which Adhikari and colleagues worked, it was postulated that people's experience of the disaster would elevate their awareness of the contribution their risk beliefs and coping capabilities could make to their capabilities, and these combined with the social interpretive variables in the CET (especially when researching a collectivistic culture) created a composite model. Combining the latter variables from PMT with CET variables demonstrated that a composite theory can work (Adhikari et al., 2018).

Drawing on the theoretical and empirical work on preparedness discussed in this and the preceding chapter, a tentative matrix is described (Figure 4.1). In addition to including the kind of multi-level approach advocated by proponents of adaptive governance (see above), this matrix differentiates factors depending on whether they constrain or facilitate preparedness. There are other issues that make adopting the kind of model depicted in Figure 4.1 worth considering.

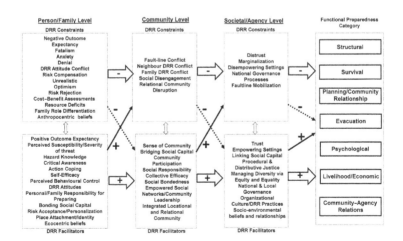

Figure 4.1 A matrix model of person, community and societal preparedness predictors

Figure 4.1 includes a broader range of variables and encompasses those that constrain and those that facilitate preparedness. The broad arrows between each of the person/family, community and societal factors indicate that constraints and facilitators can co-exist at each level. The arrows linking the various boxes provide a basic overview of the complex interrelationships between variables and levels of analysis that exist at any point in time.

For instance, it is possible to envisage a situation in which person/family constraints interact with community- and societal-level constraints (Figure 4.1) to systematically undermine the likelihood of preparing. Another scenario, in which person/family constraints interact with community- and societal-level facilitators would enable higher levels of preparedness, but less than would be expected from person/family and community facilitators interacting. The theories introduced above do not offer this level of planning and assessment flexibility.

The point being made here is that the social and societal contexts in which preparedness interventions occur will comprise a mix of the several elements described in Figure 4.1. Creating a more comprehensive matrix model makes it easier to appreciate the diverse range of positive and negative factors that must be considered in assessment, planning and intervention contexts. This approach is consistent with the adaptive governance principles introduced above.

Furthermore, the fact that the majority of the factors in Figure 4.1 can be assessed using validated measures increases its utility in DRR governance and preparedness policy and in practice contexts. Describing the preparedness process in this way also makes it amenable to including community development approaches to facilitate more cost-effective community-based preparedness policies and strategies (the latter are described in more detail in Table 7.1 and in Chapter 9). Future work should be directed to assessing the feasibility of adopting this approach.

4.8 Conclusion

Preparedness plays an important role in comprehensive DRR. This chapter discussed why preparedness typically involves people adopting low-cost items, outlined strategies to enable people's ability to transition to more comprehensive (e.g., structural, community relationship, community-agency) preparedness and discussed several theories capable of supporting the development of comprehensive all-hazards preparedness. Recognition of the complex, evolving and dynamic web of information, competence and relationship influences on preparedness led to a discussion of how adaptive governance can play a role in consolidating preparedness processes within DRR. Another question asked by the SFDRR concerned the development of processes that could support international collaboration and learning. The next chapter explores this issue.

5 DRR in international contexts: cross-cultural issues

5.1 DRR in international contexts

> Cultural differences should not separate us from each other, but rather cultural diversity brings a collective strength that can benefit all of humanity ... (Robert Alan Arthur (1922–1978))

Natural processes, such as earthquakes, wildfires, floods and hurricanes, from which disasters emanate, occur the world over. However, disasters and the impacts they create are not evenly distributed around the world, with some 60 per cent occurring in Asia and the remainder being distributed throughout North and Central America, Africa, South America, Europe and Oceania. The distribution of disasters, and especially their concentration in Asia, introduces the need to accommodate cultural diversity when considering the international applicability of DRR processes.

The SFDRR calls for the development of international mechanisms capable of delivering strategic advice, coordination, partnerships and capacities for shared DRR learning across national borders (UNDRR, 2015). To pursue this goal, it is necessary to first develop frameworks capable of supporting international collaboration in a world characterized by cultural diversity (Eiser et al., 2012).

The quote by American screenwriter, film director and film producer Arthur that opens this chapter draws attention to two relevant issues. The first relates to whether cultural differences introduce a degree of separation into DRR when it is applied in different countries. That is, do differences between cultures exist and do these differences have DRR implications? The second part of Arthur's quote calls for inquiry into whether cultural diversity can bring to DRR a collective strength capable of benefiting all humanity. This chapter discusses how significant points of separation do exist between countries and also how understanding the nature of these points of separation can create a collective benefit for all humanity. First, how do countries and cultures differ and why does this matter for DRR?

Most DRR knowledge and theory has been generated in Western, predominantly culturally individualistic (Hofstede, 2001) countries (e.g., USA, Australia). However, most disasters occur in Asian countries that sit more towards the collectivistic end of the cultural spectrum. Hence, the socio-cultural settings in which theories were developed and those prevailing in the regions where most disasters occur differ substantially; they occupy different relative positions on several cultural dimensions. In relation to Arthur's quote, this not only describes how culture can separate countries from each other but also raises a need to consider what this means from a DRR perspective. This topic is introduced in the next section.

5.2 Cultural influences on DRR processes and outcomes

Cultural beliefs and practices weave their influence into all aspects of life (Matsumoto & Juang, 2008), and DRR is not exempt from them (Paton et al., 2013). This is illustrated here using examples of how cultural beliefs influence people's interpretation of the natural processes from which hazards emanate, the risks they face, and the actions they are willing to take to reduce their risk. In doing so, it provides examples of what Arthur alluded to regarding how cultural characteristics can introduce points of separation between cultures.

For instance, Gregg et al.'s (2008) research on volcanic risk management in Hawai'i demonstrated how Indigenous Hawai'ians' reverence for Pele, the goddess of volcanoes, influenced the meaning they attributed to both volcanic hazards as sources of risk and to DRR mitigation activities that involved interfering with lava flows. For Indigenous Hawai'ians, lava flows are revered rather than being perceived as sources of threat, and these beliefs extend to how they interpret certain mitigation measures.

Amongst Indigenous Hawai'ians', risk reduction strategies that interfere with lava flows are interpreted as being disrespectful of Pele and suggestions for their use generate strong negative attitudes and overt hostility towards both engineering mitigation options and the agencies suggesting these strategies. Gregg et al. (2008) assessed the strengths of these beliefs by comparing levels of support for two engineered mitigation options, building walls to divert lava flows and bombing lava flows, in residents identifying with Hawai'ian ethnicity with those identifying with other ethnicities.

Objections to using each engineering solution (building walls, bombing) were voiced by 64 per cent and 70 per cent of Hawai'ians, respectively, compared with 49 per cent and 48 per cent of those identifying with other ethnic origins (Gregg et al., 2008). Furthermore, residents who did not identify with Hawai'ian ethnicity objected more in terms of the cost and the perceived ineffectiveness of engineering mitigation rather than from their having any socio-cultural objections. Hence, the differences between these groups were more pronounced than is indicated by the data per se.

This example illustrates how risk reduction practices in culturally diverse populations can introduce points of separation between cultures. In multicultural settings, prevailing cultural beliefs can introduce differences regarding how motivation to support or reject mitigation and preparedness options is distributed throughout a population, introduce sources of conflict into communities (see Faultline Theory, Chapter 4), and reduce trust in risk management authorities. These outcomes make future DRR planning more complex and challenging. However, the opposite can also occur – cultural beliefs can provide people with capacities and capabilities that enhance DRR effectiveness.

For instance, Jang and LaMendola (2006) discussed how the Hakka Spirit, an implicit cultural characteristic of the Hakka people in Taiwan, represents a source of DRR capacities that facilitates people's ability to thrive when faced with challenging circumstances. The Hakka Spirit represents a set of culturally implicit beliefs that embody *being able to hold on firmly despite extreme adversity* and to *keep on doing something without regard to one's own strength* (Jang & LaMendola, 2006). Its constituent components are: frugality (saving for future times of shortage; cf., accumulating resources to support survival preparedness), diligence (being attentive to environmental conditions, cf., risk assessment and environmental awareness), self-reliance (continuing to function without external help), responsibility (adopting reciprocal social responsibility in tasks, cf., shared responsibility and collective efficacy), and persistence (keeping at a task until it is completed despite any challenges encountered in ways that predispose people to sustain response and recovery activities over prolonged periods of time).

An interesting facet of the Hakka Spirit derives from how centuries of collaborative responses to typhoon damage to fruit buds in agricultural areas has generated what are akin to outcome expectancy beliefs. Once a typhoon has passed, farmers get together, go round each farm in turn, and tie the buds back onto the trees. This practice morphed into outcome expectancy beliefs that also apply to earthquakes (Jang & LaMendola, 2006). The Hakka Spirit thus represents a resource that enables people to thrive in the face of challenge and change. Examples of culture-specific practices have also been recorded in Japan.

For example, Chonaikai, a form of community governance in Japan, can support citizen participation and empowerment in DRR contexts (Bajek et al., 2008; Bhandari et al., 2010). Bhandari and colleagues discuss how Chonaikai support for the Danjiri Matsuri ritual in Kishiwada City, Osaka functioned to enhance neighbourhood relations, people's sense of community, their social capital, and their trust in Chonaikai. This increased trust, according to Bhandari and colleagues, translated into enhancing both people's hazard awareness and their earthquake preparedness.

Another Japanese culture-specific process, Machizukuri (community-led place-making with care), by emphasizing neighbourhood defined by social networks rather than urban design goals, supports residents' values, social inclusivity and sense of place, and provides a context in which residents, city planners and emergency managers can collaborate to facilitate locally relevant risk reduction and recovery capabilities (Mamula-Seadon, 2018). The effectiveness of Machizukuri in this context is evident in the process being adapted in Korea (*Maeul-Mandeulgi*) and Taiwan (*She Qu-Ying Zao*) (Mamula-Seadon, 2018). So far, this discussion has focused on cultural points of separation. In the next section, discussion shifts to how cultural diversity can create collective benefits for all humanity.

5.3 Cultural diversity bringing a collective strength for all humanity

The starting point for investigating the relationship between cultural diversity and collective benefit begins by extending analysis to comparing surface- and deep-structure facets of cultural processes (Matsumoto & Juang, 2008). Doing so introduces potential points of similarity between cultural constructs in different countries.

For example, within the Hakka Spirit and Machizukuri processes introduced above it is possible to discern functions such as community participation, collective efficacy (people collaborating on local activities), and place attachment that have been identified as relevant in Western populations and settings (see Chapter 4). Thus, while the surface characteristics of the Hakka Spirit and Machizukuri are quite different from what is found in Western cultures, their deeper-structural characteristics that represent the deeper functions served by these processes have the potential to introduce points of similarity between these cultural constructs in oriental and occidental contexts and between them and the constituent variables evident in several of the theories introduced in Chapter 4.

Finding points of convergence in cultural constructs and DRR theories opens the way to exploring whether DRR theories could provide a foundation for developing the standardized templates that the SFDRR identified as being necessary to enabling international collaboration and shared learning capabilities (UNDRR, 2015). This issue is discussed next. In so doing, it illustrates Arthur's point regarding how cultural diversity can create a collective strength that can benefit all of humanity. First, it is necessary to ask whether existing (Western) DRR theories and models have comparable levels of applicability in Asian and predominantly collectivistic cultures (Eiser et al., 2012).

5.3.1 Cultural diversity and DRR

Countries differ with regard to their relative positions on several cultural dimensions (Matsumoto & Juang, 2008). For this chapter, Hofstede's (2001) model of cultural dimensions is adopted. While only one of several models, the Hofstede framework illustrates several cultural dimensions with DRR implications. What are these dimensions?

Following research in some 72 countries, Hofstede identified how culture could be described in terms of the relative position of a country in five dimensions. These are individualism–collectivism, power distance, uncertainty avoidance, masculinity–femininity, and long-term orientation.

The individualism–collectivism (IC) cultural dimension describes the relative extent to which culture promotes, facilitates and sustains the needs and goals of autonomous individuals over those of the group (collective). Power distance (PD) assesses the degree to which people expect and accept that power is distributed unequally within society and they accept and comply with the authority of those perceived as their social superiors. The uncertainty avoidance (UA) dimension captures the extent to which members of a culture feel threatened by uncertain, unknown or ambiguous situations. Regarding the masculinity–femininity (MF) dimension, a society is deemed masculine if social sex roles are clearly separated. Finally, long-term orientation (LTO) is characterized by perseverance and sensitivity to status and where thinking and action has a strong, long-term, future-orientated perspective. These dimensions have implications for several DRR processes, including preparedness, governance and recovery.

For example, the theories introduced in Chapter 4 include various combinations of person- and social-level variables. It thus becomes possible to inquire whether the relative position of a country on the IC dimension could affect the relative value of preparedness theories. Similar issues arise when exploring the cross-cultural implications of governance.

Disaster governance encompasses the policies, procedures, practices, norms and relationships developed, institutionalized and applied by organizational actors in DRR settings to accomplish societal DRR goals (Berkes, 2021; Djalante et al., 2011; Munene et al., 2018). The inclusion of norms and relationships in this definition introduces a need to consider whether cultural characteristics have governance implications. For example, the relative position of a country on the IC, PD and UA cultural dimensions will influence how norms are developed and will affect how collaborative, multi-level stakeholder relationships are developed, enacted and maintained.

For example, the relative position of a country on the IC and PD dimensions will affect the social status dynamics developed through multi-level engagement. In countries scoring high on PD, multi-level relationships will be defined more by hierarchical compliance compared with low PD countries in which higher levels of participatory and democratic decision-making will permeate governance development and implementation. Similarly, the relative positions of countries on the UA dimension would affect levels of automatic compliance with edicts from authorities. These ideas should, however, be regarded as tentative until systematically tested.

While this discussion may be seen as driving a DRR wedge between countries, a different perspective emerges when considering that norms, social status dynamics and multi-level relationships occur in all countries. That is, the relative position of a country describes a surface characteristic but the deeper-level analysis indicates that they all develop norms and social hierarchies.

Hence, the application of a surface- versus deep-structure interpretation illustrates how diverse socio-cultural processes can share comparable deep-structure characteristics. Consequently, there exist grounds for not unequivocally assuming that theories, processes and constructs developed in predominantly Western settings will or will not apply in Eastern countries. Identifying whether some level of cultural equivalence exists in DRR processes is explored next.

5.4 Cross-cultural DRR

This section introduces research examining the applicability of a preparedness theory across hazards and cultures using the IC dimension as the basis of comparison. The IC dimension is the most commonly studied, and arguably the most important, cultural dimension (Matsumoto & Juang, 2008). This is not to entirely neglect other dimensions.

For example, some of the countries used for theory comparison (e.g., Taiwan, Indonesia) score comparably high on the PD and UA dimensions, with the opposite being the case for an individualistic country (e.g., New Zealand) used for comparison. While not representing an experimental control per se, this goes some way to controlling the influence of PD and UA on preparedness decision-making (i.e., on the dependent variable), allowing the comparison to illuminate the cross-cultural utility of a DRR theory. The theory selected for this discussion is the CET (see Chapter 4).

This chapter is not arguing that this is the only theory that could be examined in cross-cultural comparison. It was selected because it comprises both individual- (outcome expectancy) and community/social-level (community participation, collective efficacy) variables. It thus provides a starting point for studying cultural differences in individualistic versus collectivistic countries (Matsumoto & Juang, 2008). It was also selected because it has been subjected to systematic cross-cultural, all-hazards testing.

This chapter first assesses the predictive utility of the CET when tested across cultures and hazards. It then introduces the benefits that arise from distinguishing between culture-general and culture-specific processes, and discusses the implications of both for DRR planning and intervention and for realizing the UNDRR goal of supporting international collaboration and shared learning in DRR.

5.5 Assessing the cross-cultural applicability of the CET

If the UNDRR (2015) goal of facilitating international collaboration and shared learning in DRR is to be supported, theories must demonstrate their cross-cultural applicability (Eiser et al., 2012). The pursuit of this issue is important. Being able to demonstrate cross-cultural applicability in DRR processes would provide a foundation for developing robust frameworks that could be used to support international collaboration and shared learning. The latter can provide the international community with several theoretical and practical benefits.

For example, demonstrating that DRR theories play comparable roles in culturally diverse countries would enable the creation of a common foundation for collaborative learning and research across national borders. This would facilitate opportunities for international collaboration and cooperation on DRR projects and make it easier to disseminate research findings from complementary research programs more widely and irrespective of the country in which the work was undertaken.

Empirical support for the cross-cultural utility of DRR theory and practice would allow the use of theory, and practice, in developing countries that otherwise lack the resources required to conduct this research themselves. Finally, evidence of theory equivalence would increase the opportunities available to risk management agencies in different countries to access and apply (evidence-informed) risk management strategies and practices from different countries. If these benefits are to be realized, the cross-cultural applicability of DRR theories developed in Western, more individualistic countries to their Asian and more culturally collectivistic counterparts must be ascertained.

Next, this chapter summarizes empirical research on the cross-cultural equivalence using the CET (Paton, 2008). The comparative analyses were undertaken in countries situated at higher, medium and lower positions on Hofstede's (2001) IC dimension. Earthquake preparedness was examined in New Zealand (higher individualistic), Japan (intermediate collectivistic) and Taiwan (high collectivistic). Volcanic preparedness was explored in New Zealand (higher individualistic), Japan (intermediate collectivistic) and Indonesia (high collectivistic) (Jang et al., 2016; Paton et al., 2013). These analyses are summarized for earthquake and volcanic preparedness in Figure 5.1.

Comparative research applying the CET supports the view that the more people believe that people can mitigate their risk and develop their preparedness (*positive outcome expectancy*), the greater their experience collaborating with others to resolve local problems (*community participation* and *collective efficacy*), and the more they believe that their relationship with civic risk management agencies assists their ability to achieve their DRR goals and outcomes (*empowerment*), the more likely people are to *trust* risk management/scientific agencies and the information they provide and to use this information to further their preparedness outcomes (Paton et al., 2013). These analyses support the view that the CET has some measure of cross-cultural applicability. Accommodating all-hazards and cross-cultural issues in DRR models increases confidence in their applicability irrespective of the location, the hazardscape prevailing within a jurisdiction, or the socio-cultural characteristics of the

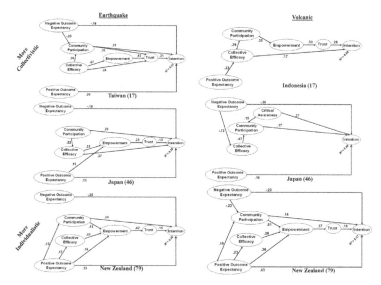

Note: The numbers in parentheses signify the IC score for each country. High scores signify greater individualism, lower scores indicate a more collectivistic country.
Source: Adapted from Bajek et al., 2008; Paton et al., 2013.

Figure 5.1   A comparative analysis of earthquake and volcanic research using the CET in countries situated at different points on Hofstede's (2001) IC dimension preparedness predictors

population being encouraged to prepare. These analyses thus depict the CET as a culture-general framework (i.e., it applies to all cultures).

However, given the importance of systematically exploring the relationship between the surface- (i.e., how a construct is expressed and applied in socio-cultural life) and deep-structure characteristics of cultural constructs, comprehensive cross-cultural testing must go further. That is, while a role for CET variables was supported in each country, this cannot be taken to assume that how people participate with others, how their collective efficacy is developed and applied, or the processes that empower community actions, is comparable in each country.

Consequently, to fully appreciate how the CET operates in practice in each of the countries depicted above, it is essential to understand how culture-specific characteristics in each country enable the theory to function in that country. The latter, and particularly how relevant culture-general and culture-specific processes complement one another, will inform understanding of how the kind of collaborative framework envisaged by the UNDRR can be operationalized and applied internationally. How this can be accomplished is illustrated by using insights from Taiwan, Japan and Indonesia. The relationship between culture-general and culture-specific processes is illustrated in Figure 5.2.

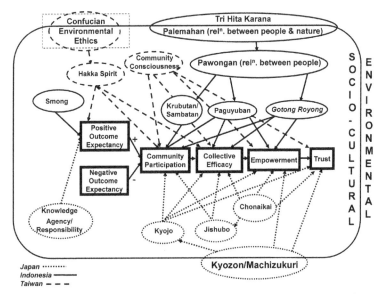

Source: Adapted from Kitagawa, 2015; Paton & Sagala, 2018; Paton, Jang & Irons, 2015.

Figure 5.2   The relationship between culture-general and culture-specific processes in Taiwan, Japan and Indonesia

For example, in Taiwan, the Hakka Spirit (see above) components of outcome expectancy, reciprocal support, collaborative problem-solving and persistence correspond to the outcome expectancy, community participation and collective efficacy constructs in the CET (Figure 5.2). In Kagoshima, Japan, culture-specific constructs such as *Kyojo* (helping each other through cooperative commitment), a construct that emphasizes the need for collaboration, cooperation and commitment in DRR, correspond to the community participation and collective efficacy constructs in the CET, and the local governance constructs, Chonaikai and Jishubo, describe mechanisms that map onto the empowerment and trust variables in CET (Bajek et al., 2008; Bhandari et al., 2010; Kitagawa, 2015) (Figure 5.2). The emergence of DRR processes in Kagoshima emphasizing the importance of being personally responsible for being able to respond proactively is analogous to positive outcome expectancy (Paton, Jang et al., 2017) (Figure 5.2). Examples of similar relationships between CET variables and socio-cultural processes in Indonesia can also be discerned and linked to the overarching influence of the Tri Hita Karana philosophy, including *gotong royong* (Paton & Sagala, 2018) (Figure 5.2).

The *gotong royong* construct generates community processes that map onto the community participation and collective efficacy components of the CET (Figure 5.2). *Gotong royong* describes community members' collective commitment to resolve day-to-day problems collaboratively, represents a medium for social empowerment via community leadership, emphasizes the maintenance of harmonious relationships, and enables reciprocal assistance when disaster strikes (Paton & Sagala, 2018). Other socio-cultural processes can also be found in Indonesia.

For instance, Krubutan (reciprocal work commitments) and Sambatan (reciprocal assistance between neighbours) which, in conjunction with Paguyuban (informal community-based organizations that support common community needs and interests), represent cultural equivalents of community participation, collective efficacy and empowerment processes (Figure 5.2). In addition, the cultural and administrative leadership construct, Musyawarah (Paton & Sagala, 2018), can be mapped onto the community empowerment facet of the CET (Figure 5.2). These relationships remain provisional until additional research into the relationships between culture-general and culture-specific constructs is undertaken.

Notwithstanding, the parallels that emerge from these comparative analyses of countries distributed along the IC dimension suggest that internationally applicable DRR theories can be developed in ways that support the SFDRR goal of making strategic advice, collaboration, coordination, learning and intervention practices available internationally (UNDRR, 2015). Doing so requires developing strategies that integrate culture-general and culture-specific processes.

These examples thus illustrate Arthur's point regarding how differences between countries can be reconciled in ways that can create a collective benefit for all humanity by opening opportunities for developing international DRR collaboration and learning processes. It is also conceivable that humanity can benefit in other ways from cross-cultural research, with one possibility arising from cultural differences in how socio-cultural-environmental relationships are conceptualized.

5.6 Socio-cultural-environmental relationships and DRR

In Chapter 1, the benefits of situating DRR processes within a socio-cultural-environmental framework were discussed. This conceptualization reflects how, ultimately, disaster results from the decisions societies and citizens make regarding how people decide (e.g., urban development, agricultural, land-use planning) to relate to a potentially hazardous natural environment.

In this context, a goal of DRR is concerned with reconciling the benefits societies and citizens accrue from their environment (e.g., fertile soils, forests, water supplies) with the co-existence strategies that enable them to eliminate or minimize their risk and develop their capacity to cope with, adapt to and learn from circumstances that arise when their environment turns hazardous. There are, however, differences between cultures regarding the extent to which environmental co-existence beliefs are accommodated in daily life in DRR conceptualizations. The next section draws on international comparisons to offer insights into how socio-cultural-environmental beliefs and relationships can be implicated in supporting how DRR goals are achieved. It is important to qualify the discussion by stating that while the socio-cultural-environmental processes discussed are not always practised as intended, they serve to introduce how socio-cultural-environmental relationships can be developed to support DRR.

For example, in Indonesia, one of the three processes that comprises the overarching socio-cultural construct, Tri Hita Karana, is the Palemahan (relationships between people and their environment) construct (Paton & Sagala, 2018) (Figure 5.2). In Japan, the *Kyozon* and Machizukuri constructs introduced earlier describe DRR activities underpinned by people's socio-environmental relationship beliefs and practices (Figure 5.2).

In Taiwan, long-held cultural beliefs regarding the importance of not fighting against or trying to beat nature, but rather living in harmony with (co-existing with) it (Paton et al., 2016) inform several DRR practices (Figure 5.2). A belief in the importance of harmony between humans and nature is a traditional facet of Confucian ethics that identifies how the cultivation and maintenance of positive relations between people and the natural world is crucial to people's quality of life (Nuyen, 2011; Yu, 2018) and to being applied to influence resilience to environmental hazards (Xiong, 2010).

This last section was introduced to illustrate how Western DRR thinking could be informed by lessons from Indigenous and Asian cultures. This process could be extended by conducting the kind of comparative analysis summarized in Figure 5.1 using a theory originating in Asia. A starting point for pursuing this possibility into practice is introduced next.

5.7 The community consciousness model

The cross-cultural comparisons discussed above started with a Western-derived theory and then assessed its applicability in several Asian countries. Another approach is to perform this process in reverse by starting with an Asian model. Some preliminary work along these lines was conducted following the 1999 '921 earthquake' in Ho-Ping township, Taiwan (Paton et al., 2016). As a consequence of being isolated by the earthquake, Ho-Ping's residents developed a form of adaptive capacity they labelled as *community consciousness* that they used to support their recovery and rebuilding activities. The components that emerged were community consciousness (community beliefs in their capacity to respond, strengthening community–environmental relations), community participation, trust, and organizational networks (cf., empowered community and empowering civic settings). This model is depicted in Figure 5.3.

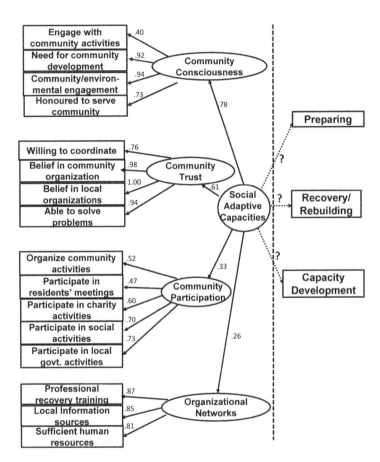

$\chi 2 = 157.01$, df = 100, p=0.137; c2/df = 1.57; RMSEA=0.046, NFI = 0.91, CFI=0.965, GFI/ AGFI=0.897/0.86). N = 172.

Source: Adapted from Paton et al., 2016.

Figure 5.3   The community consciousness model

While this model has yet to be subjected to an assessment of its cultural equivalence, there are grounds for doing so. For example, parallels can be discerned between the content of the community consciousness model and variables used in the CET comparison. The community consciousness model includes constructs that map onto the collective efficacy, community participation, empowerment, and trust variables found in Western theories such as CET and social capital theories. The cultural equivalence of this model remains tentative until it is systematically tested. Caution in conducting such comparisons is also warranted until the community consciousness model is first tested for its ability to demonstrate an ability to influence core DRR processes such as preparedness (Figure 5.3).

5.8 Conclusion

The comparative work discussed here offers insights into how mechanisms capable of delivering strategic advice, coordination, partnerships and capacities for shared DRR learning across national borders (UNDRR, 2015) could be developed and made available to support international collaboration and learning goals. This discussion also highlighted the importance of understanding the role played by culture-specific processes. This knowledge is essential for developing effective and culturally and locally relevant interventions.

The CET is only one of several theories that could have cross-cultural applicability. It was used here because it has been subject to systematic all-hazard and cross-cultural testing. Others, such as Protection Motivation Theory and Social Capital Theory, have been tested in different countries, but not yet at a level that includes either the systematic assessment of variable equivalence or theory applicability in ways that encompass all-hazard and cultural issues.

While the comparative analyses of the CET indicate an ability to support international learning and research, applying this (or any other) theory to developing a preparedness strategy requires the synthesis of culture-general and culture-specific processes (see Figure 5.2). Future research would benefit from systematically analysing constructs and theories developed in Asian DRR and inquiring whether the constructs developed in Western theories map onto them (e.g., Figure 5.3).

Additional work is needed to systematically investigate the DRR implications of cultural dimensions such as PD, UA and LTO (e.g., do they influence acceptance of top-down processes, focus on rules-based activities, and affect levels of persistence and sustained adoption of DRR capabilities?) and extend work to include other cultural dimension frameworks such as that developed by Trompenaar (Moon & Choi, 2001), to add depth to endeavours directed to assessing the international applicability of DRR processes.

The work summarized here suggests that cross-cultural comparative research can develop frameworks capable of supporting collaborative, cooperative and shared learning approaches in ways that transcend national borders. Confidence in the cross-cultural applicability of DRR theories, such as that demonstrated here (Figures 5.1 and 5.2), would provide countries that lack the resources to research resilience themselves with a foundation for their risk management planning and provide opportunities for collaborative research. It remains ever important that culture-specific factors are not ignored.

The work of Gregg et al. (2008) highlighted how DRR processes and the cooperative and collaborative stakeholder relationships it relies on can be disrupted by neglect of cultural beliefs and practices. A failure to accommodate cultural beliefs resulted in sections of the community rejecting proposed mitigation options, increased (multicultural) community fragmentation, and lessened trust in risk management authorities. DRR actions that conflict with cultural beliefs or practices are inconsistent with social justice principles, and this becomes an important issue when developing DRR processes in multicultural countries such as Australia and the USA.

In increasingly pluralistic societies, risk management strategies must be applied against a backdrop of growing cultural diversity. Reconciling this diversity with the costs and benefits of DRR strategies in fair and just ways is a demanding task and one that requires these strategies to be based on social justice principles. A relevant issue here is that the challenges to developing socially just DRR strategies in multicultural countries extend beyond issues of cultural diversity per se. To be consistent with procedural and distributive justice principles, strategies must accommodate culture distance (the relative cultural positions of migrant versus home country citizens on several cultural dimensions), the quality of acculturation over time, and how high-stress events such as disasters activate culture-specific beliefs and behaviours migrants use to respond in planning and intervention design.

This issue is most significant when host and migrant citizens occupy divergent positions on cultural dimensions (e.g., Asian citizens migrating to highly individualistic countries such as Australia). Accommodating these issues in planning starts with understanding how cultural characteristics can affect DRR planning and intervention, and the content of this chapter can inform the development of research programs exploring the multicultural implications of DRR.

Another facet of DRR that is not constrained by national borders stems from the fact that, when disasters occur, people are faced with the challenges of responding to their impacts and recovering from consequences they create. This is the subject of the next chapter.

6   DRR in response and recovery settings

> By three methods we may learn wisdom: First by reflection, which is the noblest; second by invitation, which is easiest; and third by experience, which is the bitterest.
>
> Confucius (551–479 BCE)

6.1   DRR in response and recovery settings

Chapters 2 and 3 introduced the unfortunate reality that faces contemporary DRR; many people decide not to prepare for disaster or do so in only rudimentary ways. Many people elect not to learn by either reflection (see Chapter 2's discussion of anticipation) or by invitation (e.g., from risk management agencies). Instead, they opt for Confucius' third option and put themselves in a position in which they must learn from experience. It does, however, seem that learning can occur.

When disaster strikes, people who are not prepared or who are underprepared will find themselves having to react to their circumstances. In many cases, however, they react in ways that culminate in their developing an ability to cope with and adapt to the circumstances in which they find themselves.

While not detracting from the importance of pre-event preparedness, investigating how ill-prepared and underprepared people react to the disaster consequences they encounter and develop in situ ways to cope and adapt could inform the development of strategies to support people people's ability to deal with current disasters. Work of this nature can have additional benefits, including informing how pre-event preparedness could be developed.

Current approaches to preparedness pay scant attention to the complex, dynamic and sometimes enduring circumstances that people encounter in disaster recovery settings. Some of these circumstances derive from the natural processes themselves. For instance, earthquake aftershock sequences and some volcanic events can result in people having to (re)cycle through response and recovery processes repeatedly and over prolonged periods of time. The 2011 Tōhoku earthquake in Japan had an initial magnitude of 9.0 and was followed by over thirty aftershocks with magnitudes over 6.0. The Christchurch earthquake presented the city's inhabitants with an aftershock sequence that resulted in their having to deal repeatedly with seismic challenges over many months (Becker et al., 2019).

Other recovery challenges can be traced to what Quarantelli (1996) labelled *response-generated demands* (e.g., coping with relocation, dealing with government recovery agencies, negotiating with insurance and building trade issues). The latter can permeate recovery and rebuilding contexts for months or years (Nakagawa & Shaw, 2004; Paton et al., 2014). In many cases, the challenges people face represent a mix of enduring hazard effects, a lack of resources and supportive relationships, and response-generated demands. Knowledge acquired from researching people's response and recovery experiences can inform the development of more inclusive preparedness strategies, provide survivors with guidelines for managing the diverse recovery challenges they will encounter, and offer input into BBB or capacity development programs (see Chapter 8).

Researching people's recovery experiences can provide valuable insights into how, even when affected by the same natural event (e.g., the same earthquake), people's experience is varied and dynamic. For example, the specific mix of event, demographic, social, cultural, societal, and environmental characteristics tend to be unique to a specific area and will vary from one neighbourhood to another and over time. Knowledge of this diversity can inform how local preparedness strategies are developed, contribute to the design of exercises and drills (see Chapter 7), and provide government and NGO response agencies with more detailed inputs into their planning, training, exercising and response practices.

This chapter contributes to this discussion from two perspectives. The first summarizes the application of constructs known to facilitate coping and adaptation to challenging events. The second perspective introduces comparative studies that seek to model the recovery experience and how people rise to the challenges encountered.

6.2 Applying coping and adaptation constructs to recovery

Several theories and constructs with established track records in facilitating coping and adaptation have been researched in disaster recovery settings. For example, bonding social capital has been implicated in facilitating independent (of societal resources) recovery and the mobilization of community action to support the needs of disaster victims (Hawkins & Maurer, 2010). Nakagawa and Shaw (2004) discussed how social capital influenced not only rates of disaster recovery, but also people's level of satisfaction with planning (e.g., land-use planning) processes within recovery settings. In doing so, Nakagawa and Shaw highlighted the pivotal roles community leaders play in the application of social capital within recovery settings, particularly regarding leader influence on collective decision-making and actions that support effective community recovery (McAllan et al., 2011; Williams et al., 2021).

Other social capital research has highlighted how accommodating community diversity in recovery processes is crucial for ensuring procedural distributive justice principles apply in recovery settings. For example, developing networks and building trust in some sub-groups can exclude others from recovery processes, restrict access of some groups to recovery resources, and sustain or magnify pre-existing social disadvantage and fuel social fragmentation (Elliott et al., 2010; Williams et al., 2021). Williams and colleagues observed that if residents believe that others become involved in volunteer work for their own benefit rather than the collective good, trust in recovery groups and the external agencies that support them is eroded, with trust being replaced with social disengagement and community fragmentation. They also discuss how perceived inequality in both the process of tendering for competitive community grants and grant award outcomes results in higher levels of social fragmentation between groups within a town.

It is important that community groups, and the government and NGO agencies involved in supporting them, involved in long-term rebuilding activities are aware of the potential for such outcomes to arise and develop assessment and management strategies to either prevent social fragmentation or to include conflict management strategies that can manage this outcome. Well-conceived conflict management strategies can address fragmentation issues and can, potentially, use such sources of diversity for strengthening social cohesion over time (Paton & Buergelt, 2019).

These observations reiterate the importance of social recovery strategies being based on procedural and distributive justice principles (Guion et al., 2007; Reininger et al., 2013). In doing so, it can be postulated that recovery and rebuilding strategies can profit from fostering the expression of gratitude as a way of contributing to distributive justice outcomes.

Andreason (2007) and Raggio and Folse (2011) discuss the benefits accruing from people's participation in response, recovery and rebuilding strategies including formal and informal expressions of gratitude directed to those involved. Including expressions of gratitude in recovery communications can foster the emergence of future prosocial behaviours through providing moral reinforcement and creating contexts which develop and sustain social cohesion and sense of community (Andreason, 2007; Williams et al., 2021).

Public expressions of gratitude have also been linked to increasing the quality of external social and economic assistance offered to affected communities and with enabling the continued development of prosocial behaviours beyond the end of the recovery phase (Jang & LaMendola, 2006; Raggio & Folse, 2011; Williams et al., 2021). Other psychological constructs and theories may have a role to play in this context.

For example, Drury (2012) and Sullivan and Sagala (2020) discuss how people's post-event social identities can contribute to developing a collective resilience capability that mobilizes social support and empowers collective action in recovery settings. Drury's finding that trust mediates the relationship between shared social identity and empowerment raises the prospect of exploring how it can be integrated with theories (e.g., CET, social capital) in which trust is implicated (see Chapter 4).

Other research has explored how disaster recovery can be supported by providing access to media that can inform people about experiences that could assist recovery. Song lyrics provide an example of the latter

(Paton et al., 2022). Song lyrics can capture recovery issues ranging from warnings, the implications of being ill-prepared, the physical and emotional impacts of disaster, the value of helping others, the implications of fatalism and anxiety, and how resilient and positive outcomes can emerge from people's disaster experience.

The benefits of exploring this line of research can be traced to finding that songs have been implicated in enabling people's capacity to cope with the physical and emotional aspects of their recovery experience, facilitate social bonding, develop people's sense of community and place attachment, and facilitate actions by increasing people's appreciation of their shared disaster experience (Paton et al., 2022). Future work is needed to explore how the information provision and social capacity functions of songs could be developed to support recovery practices.

An interesting aspect of song lyric analyses is their identifying how people (e.g., musicians, composers) who are unconnected with DRR but who, it can be inferred, are careful observers of people, can capture crucial facets of people's disaster experience (e.g., the consequences of ignoring warnings, how disasters facilitate social bonding). Song lyrics are not the only social route to exploring lay understanding of disaster experiences and their implications. Disaster survivors themselves represent a resource capable of providing comparable, if not deeper, insights into disaster recovery experiences. This issue is explored in the next section.

6.3 Survivor perspectives on response and recovery demands and adaptive capacities

While acknowledging that every survivor, every neighbourhood, and every group of survivors will experience a disaster differently, it remains feasible to identify common denominators across events. What this means for understanding disaster recovery is the topic addressed in this section.

If common denominators can be discerned in people's accounts of their disaster experience, this knowledge can be used to inform, but not prescribe, the development of social recovery and rebuilding models that can support intervention planning. Such models can also inform the development of research questions and hypotheses that can be applied to future recovery research. To pursue this idea, two lines of inquiry suggest themselves.

The first concerns exploring the relationship between hazard consequences and the demands people must cope with and adapt to. The second focuses on the resources, sourced from personal, family, community and societal sources, which citizens and societies can call upon to help them cope with and adapt to recovery and rebuilding demands. Key facets of both of these are illustrated in Figures 6.1 and 6.2. The content of these figures will be progressively introduced below.

Regarding the consequences and the demands people must cope with and adapt to, these are several and include those emanating from direct (e.g., earthquakes and aftershocks), cascading (e.g., landslides triggered by earthquakes or intense rainfall) and secondary (e.g., loss of lifelines like water and sewerage services, landslides) sources. These events can extend over prolonged periods of time (e.g., aftershock sequences), as do the response, recovery and rebuilding challenges they create for people.

The response, recovery and rebuilding challenges people face also come from the formal intervention and reconstruction processes (e.g., dealing with government agencies, insurance companies, builders) they will experience over periods of weeks, months or years (Nakagawa & Shaw, 2004; Paton et al., 2014; Quarantelli, 1996). These event- and response-generated demands encapsulate the adaptive demands that disaster response, recovery and rebuilding will present to survivors and describe what survivors must cope with and adapt to. The illustrative examples of adaptive demands in Figures 6.1 and 6.2 were constructed from analyses of two events, the 2011 Christchurch earthquake (New Zealand) and the Chi Chi (921) earthquake (Taiwan) in 1999, respectively (Paton et al., 2014; Paton, Jang & Irons, 2015).

In New Zealand, data were sourced from thematic analysis of life course focus group interviews with residents from five affected suburbs and 21 individual interviews (Paton et al., 2014) in August 2011 and captured residents' recovery experiences between February 2011 and August 2011, including experience of significant aftershock events. The use of the life course interviewing approach (Hutchison, 2005) facilitated capturing the time course of people's experiences over several months in ways that supported identifying the progressive development of their adaptive capacities and their application to the adaptive demands encountered (Figure 6.1). Data from Taiwan, sourced three years after the 1999 earthquake, provided insights into people's lived recovery experiences as they dealt with the continuing aftershock sequence. The findings are summarized in Figure 6.2. Including examples from culturally diverse countries increases the international applicability of these findings. However, at this stage, this framework should be viewed as a guide rather than it being prescriptive.

6.3.1 Recovery and rebuilding challenges

The models depicted in Figures 6.1 and 6.2 comprise two sections. The boxes below the 'time line' summarize the adaptive demands people identified having to cope with and adapt to in the *immediate impact*, *impact*, *response* and *recovery* phases of the respective disasters. In both cases, data were collected against a backdrop of continuing aftershock activity. Figures 6.1 and 6.2 illustrate how the sources and types of challenges people experience unfold over time, with the content of each box representing different *adaptive demands*. The series of ellipses above the time line summarize people's accounts of the personal, social and environmental resources they identified as using to confront these adaptive demands and how their application varied from one adaptive demand (boxed below time line) to another (see below).

Figures 6.1 and 6.2 summarize how these resources emerged from various combinations of personal, family, community, societal and environmental factors, with these making interdependent contributions to recovery outcomes in both cases. The construct labels and solid arrows linking core constructs (e.g., place attachment, leadership, collective efficacy) describe how they relate to each other, how they function across levels of analysis, and depict resources (e.g., trust, leadership, empowerment) that facilitate social cohesion in recovery settings (Hawkins & Maurer, 2010; Irons & Paton, 2017; Monteil et al., 2020; Williams et al., 2021). The dotted lines/arrows depict people's generalized accounts of how they applied their adaptive capacities to managing the impact, response and recovery demands encountered. These cases are discussed in the next two sections, starting with Christchurch.

6.3.2 Christchurch

In Christchurch, people reflected on the inadequacies in their structural readiness (e.g., not securing homes to their foundations), particularly regarding how it made their recovery more challenging than it needed to be (e.g., better structural preparedness could have negated needing to temporarily evacuate or reduced the level of repairs needed to make one's home habitable). Their reflections prompted their recognizing the important roles structural measures would play in their future preparedness. The implications of this issue are recorded in the *immediate impact* box in Figure 6.1.

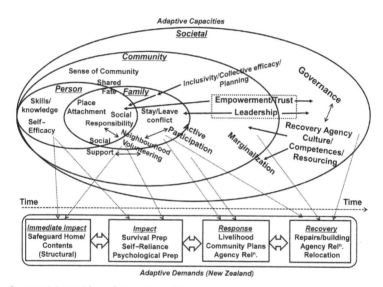

Source: Adapted from Paton et al., 2014.

Figure 6.1   A summary of the interdependent influences of person, family, community and societal adaptive capacities on individual and collective response to adaptive demands experienced following the 2011 Christchurch earthquake

Potentially preventable demands were also attributed to limitations in survival (e.g., only having one day's supply if water, no torch or having a torch but no batteries) and family preparedness. For example, a lack of stored water and food and alternative cooking sources meant their having to divert time and energy to accessing essentials rather than dealing with family needs, house repairs and employment issues. A need to enhance personal and family self-reliance was thus seen as an adaptive demand, and one that was seen by many as a future priority. A significant demand emerged from people identifying their need to develop strategies to manage stress in themselves and their family members (see impact box in Figure 6.1). A majority of Christchurch respondents identified psychological preparedness as being as important to their future preparedness as structural preparedness.

As residents advanced their abilities to manage impact issues by securing survival essentials, they transitioned into the response phase (Figure 6.1). The key adaptive demand at this time derived from responding to rectify the losses and damage experienced in their neighbourhood in the absence of formal, external support.

As the activities of these self-help neighbourhood groups brought stability to respondents' recovery context, people were in a better position to turn their attention to dealing with other prominent adaptive demands, including responding to employment and livelihood issues (Figure 6.1). Finally, and particularly during the recovery–rebuilding transition, societal response issues (cf., response-generated demands) became more prominent adaptive demands. These included demands from dealing with builders, tradespeople and insurance companies. Other demands identified as significant at this stage included dealing with temporary (and potentially permanent) relocation. Several of these issues are reiterated in the second study of recovery from Taiwan.

6.3.3 Taiwan

In Taiwan, survivors faced similar adaptive demands to their New Zealand counterparts (Figure 6.2), though some differences were present. Taiwanese survivors echoed the views of their New Zealand counterparts in emphasizing how their inadequate pre-event structural preparedness created potentially avoidable problems, with the issues arising being described as a major 'immediate impact' demand (Figure 6.2) and one that required work on potentially preventable activities such as executing housing repairs. A point of departure in the Taiwan analysis was livelihood and economic challenges being cited as 'immediate impact' demands. This reflects the affected area being an agricultural one, with demands on livelihood and the local economy arising immediately as a result of crop damage and infrastructure losses (e.g., being unable to transport fresh produce to markets).

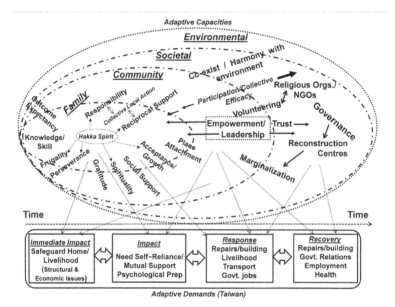

Source: Paton, Jang & Irons, 2015; Paton et al., 2016.

Figure 6.2 A summary of the interdependent influences of person, family, community and societal adaptive capacities on individual and collective response to adaptive demands experienced following the 921 (Chi Chi) earthquake in Taiwan in 1999

Like their New Zealand counterparts, Taiwanese respondents identified stress as a major demand (Figure 6.2), and one they were similarly ill-equipped to manage. A need to develop collaborative approaches to dealing with local issues was similarly identified as a 'response' demand. Livelihood issues continued to be described as significant issues in this period. Another difference between the cases was that in Taiwan, *government jobs* was cited as a response demand (Figure 6.2). This reflected how disaster governance in Taiwan includes establishing government supported reconstruction centres in disaster-affected areas.

Reconstruction centres provide training and support for establishing new employment opportunities in disaster-affected communities. However, the employment opportunities supported by reconstruction centres did not always map onto survivors' previous jobs, and this became problematic if available employment options were of lower status than those previously enjoyed by community members. Hence, people's engagement in reconstruction centres could either support or marginalize survivors (Figure 6.2).

As in New Zealand, relationships with government agencies and mobilizing collaborative activities to deal with recovery challenges (e.g., repairing homes, meeting the needs of more vulnerable members of the community) were identified as recovery issues. In Taiwan, dealing with health issues was also identified as sitting in this category (Figure 6.2).

It was also evident, in both New Zealand and Taiwan, that people were able to adapt to these atypical circumstances. Interviews with residents provided insights into the range of personal, community and societal factors implicated in accounting for these differences. Participants described several ways in which adaptive capacities at person, family, community and societal levels of analysis they had access to or developed facilitated their being able to cope and adapt to the circumstances in which they found themselves (Figures 6.1 and 6.2).

6.4 Adaptive capacities in Christchurch

In Christchurch, residents attributed their ability to cope with the immediate impact and impact challenges involved their being able to apply relevant knowledge and skills (e.g., DIY, trade skills) and having a 'mind-set' of 'getting the job done' (cf., self-efficacy) (Figure 6.1). In several cases, these skills formed the basis of adaptive strategies involving people developing functional collaborative and cooperative relationships with neighbours to form organized neighbourhood self-help groups to deal with various local demands (e.g., removing rubble from damaged buildings, helping repair others' houses, establishing community meeting places, looking after more vulnerable residents). This is reflected in the role of volunteering (Figure 6.1).

Family characteristics such as family members collectively feeling attached to living in Christchurch (place attachment, sense of shared fate, social responsibility) and family agreement about staying in Christchurch and collaborating with others to confront local issues were cited as adaptive capacities in the immediate impact and impact phases (Figure 6.1). These factors were additionally implicated in facilitating family access to social support, with this contributing to their ability to manage their stress. In contrast, family disagreements about whether to stay in Christchurch diminished the likelihood that family members would commit to engaging in local response and recovery tasks.

Personal and family decisions to remain and to collaborate built on people's sense of shared fate with others and contributed to their feeling a sense of belonging and identity. The sense of connectedness to people and place this generated reinforced people's commitment to supporting local recovery initiatives (Paton et al., 2014) (Figure 6.1). Furthermore, inclusive and active involvement in local activities, and the availability of supportive local leadership, increased people's access to social support from a broader social network and enhanced their collective knowledge and skill base (cf., collective efficacy) in ways that created an adaptive resource that could be applied to dealing with response and recovery demands (Figure 6.1).

Trusted local leaders were identified as a valuable resource. Leaders were instrumental in sustaining community connectedness and linking community with external agencies and specialists to secure the resources and assistance required to meet local community needs and empower local recovery initiatives (Figure 6.1). The degree to which community and agency activities complemented each other at this stage was a major determinant of the quality of people's recovery experience. However, some focus group members reported feeling marginalized by government agency roles. This was attributed to perceptions of inadequate coordination and a lack of local knowledge in agency personnel (Paton et al., 2014). Relationships perceived as marginalizing or disempowering reduce trust in agencies and diminish the quality of future relationships (Figure 6.1).

6.5 Adaptive capacities in Taiwan

Some similarities and differences emerged in the adaptive capacities found in Taiwan. Prominent differences included the family, rather than the individual, being the base unit of analysis. Another significant difference was the important role the inclusion of an *environmental* level of analysis played as an adaptive capacity in Taiwan (Figure 6.2). Differences were also evident regarding volunteer contributions. This was identified as coming from two sources. One was from within community settings. This is reflected in how people's beliefs regarding responsibility and reciprocal support interacted to facilitate collective local actions (Figure 6.2). Volunteering in Taiwan was also influenced by NGOs and religious organizations mobilizing large numbers of volunteers (Figure 6.2) from around Taiwan to assist recovery efforts (Paton et al., 2016).

In Taiwan, the hashed ellipses around the family and community levels of analyses were applied to signify that family and community interests and capabilities are more closely interrelated than was the case in New Zealand. The left-most ellipse describes adaptive capacities sourced predominantly from the Hakka Spirit (see Chapter 5). The characteristics of the Hakka Spirit (frugality, diligence, self-reliance, shared responsibility and persistence) were collectively implicated in facilitating coping and adapting to the demands encountered in the immediate impact, impact and response phases (Figure 6.2). This reflects how this adaptive cultural resource facilitates responding in reciprocal and collaborative ways as a matter of course (in New Zealand, people made conscious choices to collaborate).

The inclusion of an environmental level of analysis in Taiwan is significant (Jang & LaMendola, 2006). Implicit natural environmental co-existence practices in this population express themselves in the belief that there is always something that can done to deal with environmental hazards (see Chapter 5 and Figure 6.2). The implications of this facet of community life are discussed in more detail in Chapter 8.

Other unique adaptive capacities derived from Hakka beliefs included the importance of learning to accept disasters as part of life experience. This facilitated people's persistence in response and recovery contexts (Figure 6.2) and was reinforced by their culturally embedded spiritual beliefs. The latter help people identify inner strengths to continue in their recovery efforts over time, to find positive meanings in people's sufferings, to be optimistic about the future, and to focus on the positive aspects (e.g., people collaborating to enable the recovery of the whole community) of their experience (Jang & LaMendola, 2007; see also Figure 6.2).

Spirituality was additionally implicated in supporting people's sense of belongingness, reinforcing their social support networks, and in helping them manage psychological issues (Figure 6.2). When describing their recovery experiences, Taiwanese respondents echoed their New Zealand counterparts in singling out how community leadership and leader roles in effectively liaising with government, NGO and religious organizations (empowering recovery through joint ventures between local communities, local government, farmers, industry bodies, and humanitarian NGOs) supported their managing recovery demands (Figure 6.2).

6.6 Conclusion

This chapter described models derived from analyses of people's recovery experiences that could serve as a tentative framework for developing research questions and hypotheses for future recovery research and for guiding response and recovery planning. The accounts discussed (summarized in Figures 6.1 and 6.2) illustrate the kinds of adaptive demands that need to be prepared for and the adaptive capacities that could be the targets for such preparatory or support strategies.

In addition to reinforcing the importance of theoretical work on identifying adaptive capacity variables (e.g., outcome expectancy, self-efficacy, linking and bonding social capital, empowerment, trust), the analyses summarized in this chapter identified additional variables that could inform future theory development. Examples include social support, family conflict and family coherence, social inclusivity, procedural and distributive justice, social marginalization, social facilitation, gratitude, socio-environmental co-existence, and social responsibility. The Taiwanese findings reiterate the importance of understanding and accommodating the role that culture-specific factors (e.g., the Hakka Spirit) play as culturally implicit resilience resources capable of supporting recovery and rebuilding activities over prolonged periods of time (see Chapter 5).

The work discussed from Chapters 3 to this chapter provide evidence-informed examples of how DRR theories can be applied to achieving the DRR goal of enhancing the economic, social, health and cultural resilience of persons, communities, countries, their assets and their environment in disaster-affected contexts. The process of developing and implementing the several interventions described in this and in earlier chapters all incur costs to people (e.g., financial, social, opportunity) and society (e.g., building levees, public education). A need to consider the implications of these costs was identified in the SFDRR Priority 3 (see Chapter 1). This calls for additional emphasis to be placed on assessing whether these activities can provide a return on investment for societies and citizens. The next chapter discusses work on cost–benefit analyses of DRR interventions and introduces the role that evaluation processes can play in responding to the cost-effectiveness calls of Priority 3.

7 Assessing the effectiveness of DRR: cost-benefit and evaluation perspectives

7.1 Assessing the effectiveness of DRR

> You cannot predict the outcome of human development. All you can do is, like a farmer, create the conditions under which it will begin to flourish. (Sir Ken Robinson (1950–2020))

The Robinson quotation that opens this chapter encapsulates several issues relevant to understanding the challenges faced when evaluating DRR interventions. These issues have a bearing on the extent to which achieving a (measurable) return on investment, at least for some facets of DRR, is a realistic goal. A significant challenge to achieving this goal, as the preceding chapters introduced, is the complexity inherent in developing, integrating and applying DRR processes (e.g., governance, anticipatory planning, preparedness and response/recovery processes) in socially, culturally and environmentally diverse settings. The complementary interrelationships between these processes means that their contribution to DRR must be assessed holistically. Similar challenges exist in recovery and rebuilding settings. This complexity makes it impossible to reliably predict the outcome of DRR strategies; all that can be done, as Robinson alludes to, is to create the conditions for DRR outcomes to flourish. If it is impossible to guarantee outcomes, particularly regarding the achievement of social goals, it becomes more difficult to assess whether DRR can generate a return on investment.

However, the more tangible nature of societal-level DRR strategies (e.g., constructing levees, establishing building codes) may render them more amenable to cost–benefit analyses (CBA). Even so, it may not always be possible to remove the more subjective political and economic influences that occur in societal DRR. It is against this backdrop that work on CBA is discussed.

This chapter first reviews work on applying CBA to DRR. It then introduces evaluation processes and their potential to systematically assess the effectiveness of DRR strategies. In the next section, work reviewing CBA is presented.

7.2   Cost-benefit analyses and DRR

The complexity of DRR strategies, the diversity of the circumstances in which they must be developed and applied, and the time frames over which they must be sustained introduces significant challenges to CBA, as does a lack of systematic and comprehensive bases for designing and conducting CBA (Klima & Rueda, 2020; Kull et al., 2013; Shreve & Kelman, 2014). To the latter can be added challenges that derive from the need for CBA to encompass sources of more tangible (e.g., a levee) costs with those that entail intangible (e.g., accommodating social diversity, opportunity costs) costs and benefits. To what extent has CBA proven useful? This is the topic addressed in the next section.

Evidence exists that provides qualified support for the practical application of CBA to DRR (Khan et al., 2012; Klima & Rueda, 2020; Kull et al., 2013; Shreve & Kelman, 2014). However, these authors warn that the interpretation of CBA data must be qualified by issues relating to, for example, the complexities inherent in estimating risk and the quality of data used for CBA calculations. These authors add to this list issues regarding the potential for DRR interventions to create negative outcomes, problems assessing outcomes distributed over space and time, accommodating stakeholder demographic diversity, and a lack of protocols for integrating the perspectives of diverse community, expert and government representatives in CBA. However, some successes in CBA are evident.

Shreve and Kelman (2014) identified CBA having been applied to address economic (e.g., financial capacity to return to a previous path after a disaster), physical (e.g., susceptibility of built environment/infrastructure to damage), environmental (e.g., land and water use, ecosystem stability), and social (e.g., individual and institutional capacity to cope and respond to disaster) aspects of DRR activities. However, they also point to discrepancies in the quality of CBA outcomes across these areas.

The more qualitative and less tangible social and environmental components (e.g., evacuation planning, preparedness, ecosystem services, values attributed to recreational use of natural spaces, biodiversity, people's perception of safety, livelihood and business continuity, warning systems) fare less well than their economic and physical/structural (e.g., installing levees, cost of construction materials, maintenance, labour, time to construct) counterparts in CBA (Klima & Rueda, 2020; Shreve & Kelman, 2014). Shreve and Kelman also discuss how tangible and intangible elements can overlap. For example, while direct costs such as loss of income from disruptions to employment represent a tangible cost, the indirect contributions livelihood issues make to people's overall well-being over time (e.g., diminished social support from being unable to work) falls into the intangible category, with the latter being underrepresented in CBA data.

An important issue highlighted by Gregg and Houghton (2006) is how CBAs often culminate in planning decisions accommodating events up to a specific intensity (based on how integrating return periods and intensity interact to influence how economic and political imperatives determine the scale of mitigation actions implemented). However, estimates may not cover the full range of possible intensities, leaving potential gaps in CBA calculations. Another issue concerns CBAs adopting a restricted rather than an all-hazards focus.

Shreve and Kelman (2014) found a bias, attributed to factors such as frequency of occurrence, towards focusing on floods, droughts and earthquakes over events such as wildfires and volcanic eruptions. They continue by discussing how CBA is less likely to include cascading or ripple effects, short- and long-term recovery and relocation issues, infrastructure and lifeline impacts on recovery experiences, loss of life, injuries, mental health impacts, long-duration events (e.g., aftershock sequences, volcanic eruptions) and spatial and demographic diversity issues in CBA calculations. Added to this is a lack of consensus regarding how CBAs should be conducted (Khan et al., 2012; Klima & Rueda, 2020; Shreve & Kelman, 2014).

Overall, while there is some support for CBA and its capacity to contribute to making judgements regarding the cost-effectiveness of DRR interventions, inconsistencies in its application (e.g., tangible versus intangible costs, economic/structural versus social/ecological domains) and the lack of consensus on its conduct places limits on its current utility as a mechanism for responding to the SFDRR Priority 3 goals (Khan et al., 2012; Klima & Rueda, 2020; Shreve & Kelman, 2014). The major issue lies with the cost element of CBA. A less ambitious, but important, alternative is to start with assessing how interventions function (Klima & Rueda, 2020; Kull et al., 2013). That is, are DRR interventions effective? The next section discusses how incorporating an evaluation framework in DRR can contribute to answering this question.

7.3 Evaluation: process, content, outcome and context issues

By adding an evaluation framework in DRR planning, the inclusion of process, content, outcome and context evaluation, the goal is to attempt to make more explicit the less tangible, qualitative social and environmental elements of DRR that are currently underrepresented in CBA (Klima & Rueda, 2020; Kull et al., 2013; Shreve & Kelman, 2014). This process starts with describing process, content, outcome and context evaluation.

The goal of *process evaluation* is identifying the suite of activities and the relationships (e.g., the rationale for ordering theory variables) between the constituent components (the process) developed with the intent to enable achieving a specific outcome or impact (e.g., increasing preparedness) but not the outcome per se (cf., the separation of the independent and dependent variables in research). A second activity, *content evaluation*, describes the specific variables or components included in the process, the rationale for their inclusion, and how they contribute to the overall process. The next level, *outcome evaluation*, assesses whether applying the selected process/content combination achieves a specific goal (e.g., increase levels of structural preparedness), including determining whether and how engaging with the program made a difference to its participants (cf., face validity, the perceived meaningfulness of the process to its recipients).

It is important to appreciate that program outcomes can be positive and negative, with these outcomes arising from extraneous sources (i.e., DRR programs are administered in real life and so subject to the diverse personal, social, societal, economic and environmental demands impinging on program participants). Assessing the presence and influence of these extraneous influences is the province of *context evaluation*. Given the contextual nature of these issues, it is up to the individual researcher or practitioner to determine what contextual factors to accommodate in their framework based on their local knowledge of external factors present at a given time or over a specific time frame.

All facets of DRR (e.g., governance, preparedness, recovery) could be evaluated. The remainder of this chapter focuses on providing examples of evaluating preparedness and recovery. Regarding the former, one approach draws together the preparedness issues discussed in Chapters 3 and 4 to present a developmental evaluation framework. A second outlines an evaluation framework for preparedness drills or exercises, and a third focuses on evaluating a theory-based intervention. The final model describes a framework for evaluating recovery outcomes. Discussion commences with an evaluation framework derived from the contents of Chapters 3 and 4.

7.4 A developmental approach to process and content evaluation

The first approach to evaluation derives from conceptualizing preparedness as a progressive developmental process comprising a series of separate but related components. The rationale for doing so can be traced to the work of Bočkarjova et al. (2009).

Using the Stages of Change or Transtheoretical Model to frame their analysis, Bočkarjova et al. (2009) described a developmental preparedness process characterized by differences in the relative salience of predictors at different stages in the process. Bočkarjova and colleagues discussed how people's transitioning from a *pre-contemplation* (cf., anticipatory thinking) to a *contemplation* stage was influenced by perceived vulnerability and response efficacy variables. In contrast, to proceed from the *contemplation* phase to the *action* phase, the most salient variables were response efficacy, information regarding the severity of hazard consequences, the costs of protective actions, and people's subjective hazard knowledge. People's progression from *action* to preparedness *maintenance* was driven by their beliefs regarding the intrinsic and extrinsic benefits of hazard preparedness. Finally, the most prominent predictors of *sustaining preparedness* were response efficacy and self-efficacy. This work provides some support to the idea of conceptualizing preparedness as a process involving the progressive development of more comprehensive levels of preparedness, with the transitions being, to a large extent, influenced by different predictor variables.

While Bočkarjova et al. (2009) focused on transitional variables, the preparedness evaluation framework proposed here describes process, content and outcome factors derived from the content covered in Chapters 3 and 4. Based on this content, the framework differentiates people who are: (i) reluctant to anticipate their future risk and preparedness needs and have decided not to prepare; (ii) starting to prepare, but in limited ways, both qualitatively (e.g., favouring survival over structural preparedness) and quantitatively (e.g., the number of items adopted); (iii) engaging in activities to increase their level of preparedness in each functional (structural, community-agency, etc.) category, or (iv) are well prepared and engaged in sustaining their preparedness. These are presented in Tables 7.1–7.4, a tentative evaluation framework applied to DRR preparedness as a developmental process.

To begin, Table 7.1 describes activities to challenge impediments to anticipation and encourage people to consider engaging with DRR processes.

Once receptive to DRR information, people respond by adopting low-cost items. Other strategies are needed to motivate comprehensive measures (Table 7.2).

As people engage with DRR processes, their sense of perceived control increases. Providing information on environmental threats can now motivate future action. Preparedness theories support intervention strategy design and implementation (Table 7.3).

ASSESSING THE EFFECTIVENESS OF DRR 95

Table 7.1 Countering anticipatory reticence/decide not-to-prepare decisions

DRR Challenge	Constraint on Process	Process/Content	Outcome Measure
Unrealistic Optimism	People believe disasters will affect others but not them, and will occur elsewhere, but not where they live.	Facilitate participatory discussions of hazard consequence/preparedness with others in their neighbourhoods/social networks. Presenting interactive local hazard maps to increase acceptance of local risk.	Searching for information from societal sources. Increased personal responsibility.
Risk Compensation	Perceived need to prepare inversely proportional to perceived environmental threat. Mitigation (e.g., levees) assumed to reduce/eliminate threat to them.	Communicate importance of agency and citizen cooperation, societal (e.g., building codes) and household (e.g., securing chimneys) actions complement one another and rely on stakeholders sharing responsibility.	Accept need for shared responsibility. Aware of complementary DRR roles.
Denial, Fatalism and Anxiety	Believing hazard events are uncontrollable manifests in denial/anxiety, reducing likelihood of preparing.	Switch focus from uncontrollable sources (e.g., earthquakes) to controllable consequences and describing how specific actions lessen risk (e.g., securing internal fittings counter effects of ground shaking).	Decreased focus on hazards as causes. Greater focus on consequences. Knowledge of risk-consequence link.
	Lay people's thinking driven by associative/affect-driven, experiential thinking, results in people interpreting events/risk information in terms of fear, dread, feeling powerless, overwhelmed, and anxious.	Anxiety managed using psychological preparedness (anticipate sources of anxiety and developing appropriate stress management strategies). Invite people to consider what more vulnerable members of society (e.g., residents in a home for the elderly) could do to prepare. If believe that those less able than themselves could prepare, more likely to prepare themselves.	Lower scores on hazard-specific anxiety. Increased acceptance of capability to adopt measures to deal with hazard event consequences.

DRR Challenge	Constraint on Process	Process/Content	Outcome Measure
Overconfident	Overconfidence from assumed knowledge, infer that experience with low-intensity events prepares them for extreme events.	Provide feedback on diverse scenarios, ask people to reflect on the source and accuracy of their judgements, personalize their risk based on this appraisal. Provide information about high-intensity events and their personal and household implications.	Differentiate own knowledge from that from external sources.
Adopt low-cost items	Enable shift to preparedness thinking.	Presenting small lists, starting with low-cost items.	Adopt low-cost survival items.

ASSESSING THE EFFECTIVENESS OF DRR 97

Table 7.2 Enabling preparedness

DRR Challenge	Constraint on Process	Process/Content	Outcome Measure
The costs and benefits of preparing for future events	Preparedness costs immediate but benefits occur at an indeterminate future time. Future events seen as abstract so action delayed until threat is tangible and benefits realizable.	Encourage people to consider implications of dealing with events with short warning and response times and include information on how survival items would not protect people, include information on time required to implement structural preparedness.	Recognize survival items do not offer protection. Acknowledge value of structural measures for resilient response. Plan structural preparedness.
	Survival and everyday objects conflated with objects used regularly.	Keep emergency kit items separate from household uses, check kit regularly. Restock regularly.	Differentiate everyday and emergency kit content. Kits maintained, stored separately and accessible.
	Assume knowing where knowledge exists is their personal knowledge.	Move from general information to personalize risk knowledge that is personally and locally relevant and tailored to person and household.	Plans/knowledge are personally/locally relevant, tailored to person and household.

Table 7.3   Facilitating comprehensive preparedness

DRR Challenge	Constraint on Process	Process/Content	Outcome Measure
Facilitating and Sustaining Comprehensive Preparedness	Enable transition from adopting low-cost (e.g., survival measures) to high-cost (structural, etc.) measures.	Progressively introduce complex/costly (e.g., structural) measures over time. Describe benefits of high-cost items (e.g., enhance family safety, improve value of the home). Present information on each functional category one at a time; explain links between each action and its protective/enabling function.	Increase knowledge of benefits of high-cost items. Increased adoption of high-cost measures.
Facilitating and Sustaining Comprehensive Preparedness (continued)	Lack of social context in which to discuss issues and create locally meaningful strategies in which shared responsibility and social norms facilitate adoption.	Facilitate development/use of neighbourhood groups, invite participants to anticipate potential event consequences. Information provided socially helps people frame their information search and interpretation to enhance its personal and local meaningfulness.	Assess with information and social engagement measures from PADM, CET, Social Capital research.

DRR Challenge	Constraint on Process	Process/Content	Outcome Measure
Facilitating and Sustaining Comprehensive Preparedness (continued)	Lack of social context in which to discuss issues and create locally meaningful strategies in which shared responsibility and social norms facilitate adoption.	Integrate risk management, community development, and community empowerment processes to enable cooperative learning. Community-based leaders support planning and developing relationships with external agencies. Regular participation increases knowledge, quality of relationships, intrinsic motivation to prepare, and develops preparedness norms which sustain motivation and access to social support resources to manage stress and anxiety.	Community development strategies implemented. Community leaders functioning. Assess using social engagement measures from PADM, CET, Social Capital and TPB research.
Facilitating and Sustaining Comprehensive Preparedness (continued)	Dealing with multi-hazard scenarios.	With multi-hazard scenarios (earthquake, wildfire in the same area), discuss one hazard (e.g., earthquakes then wildfires) at a time. Discuss similarities and differences between each re. survival, structural, etc. preparedness. Facilitate discussion in neighbourhood settings to build locally meaningful content that empowers people to develop local plans and priorities, develop local knowledge, enhance perceived control, and sustain social support resources.	Assess multi-hazard knowledge. Assess knowledge of similarities and differences across hazards for each functional category. Assess using social engagement measures from PADM, CET, Social Capital research.

DRR Challenge	Constraint on Process	Process/Content	Outcome Measure
Facilitating and Sustaining Comprehensive Preparedness (continued)	The need for psychological preparedness.	Psychological preparedness develops people's coping and adaptive capabilities. It is less about eliminating people experiencing adverse stress reactions and more about enhancing understanding of how/why they react as they do and enabling development of ways to anticipate and manage their/family/social network member stress, enhance information processing, and support decision-making.	McLennan et al.'s (2020) psychological preparedness scale.
	The need for psychological preparedness (continued).	Stress inoculation and learned resourcefulness training; support psychological preparedness and stress management.	Stress inoculation and learned resourcefulness programs/strategies developed, implemented and assessed.
Facilitating and Sustaining Comprehensive Preparedness (continued)	Continue to safeguard against cognitive biases.	Strategies to counter these were discussed earlier.	[See Chapter 3 in this book]

DRR occurs against a backdrop of infrequent events, so strategies to maintain capacities over time are needed. Strategies such as collaborative learning and mentoring support developing socially just processes that accommodate diversity (e.g., ageing populations, multicultural populations) (Table 7.4).

7.5 Outcome evaluation studies

The approach summarized in Tables 7.1–7.4 is not the only evaluation option available. In this section, two contrasting examples are presented, one covering a preparedness drill and another on evaluating a theory-based community development initiative.

7.5.1 Evaluating preparedness drills

The first example discusses how participating in a Shakeout earthquake preparedness drill in New Zealand increased levels of preparedness actions and knowledge, influenced hazard-related beliefs (e.g., reduced fatalism), and enhanced levels of home, community and workplace preparedness (Vinnell et al., 2020). Regarding levels of hazard knowledge, following participation in the Shakeout Drill, some 66 per cent of participants knew the correct action ('drop, cover, hold') to take when inside a building. In contrast, amongst non-participants, 79 per cent *did not know* what to do. However, only some 14.1 per cent knew what to do when outside. The capacity of these lessons to function when experiencing an earthquake was supported. Some 63 per cent of Shakeout participants performed the 'drop, cover, hold' response when experiencing an earthquake compared with only 20 per cent of their non-participant counterparts doing likewise. The benefits of participation extended to other aspects of preparedness.

Vinnell et al. (2020) found that some 47 per cent of drill participants increased their household preparedness and 30 per cent developed a home emergency plan compared with 20 per cent and 14 per cent, respectively, of non-drill participants. They also reported that 39 per cent of drill participants increased their workplace preparedness and 33 per cent developed workplace emergency plans, compared with 5 per cent and 3 per cent, respectively, of non-participants. However, community preparedness remained unchanged. The benefits of Shakeout partici-

Table 7.4 Facilitating sustained preparedness

DRR Challenge	Constraint on Process	Process/Content	Outcome Measure
Facilitating Sustained Preparedness	Maintain interest when events are infrequent and occur against changes in demographics, risks, etc.	Develop neighbourhood/community/ social networks and build DRR strategies around community development strategies that facilitate responding to challenge and change in everyday life. Social facilitation enhances self-efficacy (persistence, developing a wider range of plans), outcome expectancy (belief that response capabilities can be developed), and enhances interpersonal communication and relationship competencies (e.g., cooperation, collaboration, personal and social responsibility).	Assess knowledge and use social engagement measures from PADM, CET, Social Capital research.
Facilitating Sustained Preparedness (continued)	Developing supportive and enabling social-structural/ community-based entities and processes.	Ideas for developing community-based groups include the Community Board concept in New Zealand, the Community/ Neighbourhood Emergency Response Team (CERT/NERT) concept, the Jishubo concept in Japan, and from mobilizing communities of interest (e.g., church, sport, social groups) to create opportunities to explore collective strategies.	Assess community group development, functioning and continuity using social engagement measures (PADM, CET, Social Capital).

ASSESSING THE EFFECTIVENESS OF DRR 103

DRR Challenge	Constraint on Process	Process/Content	Outcome Measure
Facilitating Sustained Preparedness (continued)	Developing supportive and enabling social-structural/community-based entities and processes (continued).	Neighbourhood groups use strategies, including property assessment (review structural preparedness/implementation issues, contacts to support homeowner installation), search conferences (e.g., what worked/did not work elsewhere), design meetings (e.g., implementation options in the area), and workshops (e.g., experts discuss issues of community interest) to develop collective knowledge, ideas, plans and actions, reinforce social network bonds, preparedness norms, and community competencies that sustain DRR capability over time.	Search conference/workshop programs developed, enacted and assessed (collective plans, knowledge, actions, social bonds, norms, etc.).
Facilitating Sustained Preparedness (continued)	Developing supportive and enabling social-structural/community-based entities and processes (continued).	Sustained DRR capability enabled using community liaison committees led by elected community leaders who create structured settings for residents to collaborate in formulating and implementing ideas and empower action via their ability to authoritatively represent local needs, interests and goals to government, business and risk management agencies and ensure knowledge, support and resources provided supports local DRR initiatives.	Community liaison committees with elected leaders established and social and civic functions assessed. Assess using social engagement measures from PADM, CET, Social Capital.

DRR Challenge	Constraint on Process	Process/Content	Outcome Measure
Facilitating Sustained Preparedness (continued)	Developing supportive and enabling social-structural/ community-based entities and processes (continued).	Community engagement/empowerment processes facilitate collaborative learning/ peer tutoring approaches to developing and maintaining DRR capabilities. Knowledgeable community members act as peer tutors, mentors or advocates (assist those disempowered, give them a voice in DRR processes) for those starting at lower levels of the preparedness ladder. Selection and training required.	Develop/assess peer tutoring, collaborative learning, mentoring and advocacy programs.
Facilitating Sustained Preparedness (continued)	Developing supportive and enabling social-structural/ community-based entities and processes (continued).	Training to manage inter- and intra-group conflict, mobilize knowledge, experience and attitude diversity to co-create outcomes relevant for all. Scenario planning used to enable the collaborative use of stakeholder's expertise, beliefs, needs and goals to create shared DRR goals.	Training needs analyses conducted and selection and training programs developed, implemented and evaluated.

DRR Challenge	Constraint on Process	Process/Content	Outcome Measure
Facilitating Sustained Preparedness (continued)	Developing supportive and enabling social-structural/community-based entities and processes (continued).	Collective learning/reframing of perceptions enabled using brainstorming, the Delphi method, extended contact method (cases from other communities illustrate how interpersonal conflict emerged and was resolved), multiple classification training (shift people's thinking about others from single to multiple dimensions [e.g., not only as an environmentalist, but also as a local resident, a parent, a gardener, a community volunteer, etc.]), and cooperative learning (create community-based responses).	Social facilitation programs to develop shared responsibility and collaborative approaches to DRR developed, implemented and assessed.
Facilitating Sustained Preparedness (continued)	Developing supportive and enabling social-structural/community-based entities and processes (continued).	Sustain preparedness using community development strategies that build competencies relevant for dealing with everyday challenges and opportunities and incorporating risk management and DRR strategies in complementary ways.	Community development program operating/ implemented, functioning assessed.

Sources for Tables 7.1–7.4: Adams et al., 2017; Aldrich & Meyer, 2015; Andreason, 2007; Armaş et al., 2017; Ballantyne et al., 2000; Becker et al., 2013; Bočkarjova et al., 2009; Buergelt & Paton, 2014; Burton et al., 1993; Chaiken & Trope, 1999; Charleson et al., 2003; Cohen & Abukhalaf, 2021; Crozier et al., 2006; Daniel, 2007; DiPasquale & Glaeser, 1999; Etkin, 1999; Frandsen et al., 2012; Grothmann & Reusswig, 2006; Guion et al., 2007; Harries, 2008; Hawkins & Maurer, 2010; Johnson & Nakayachi, 2017; Kerstholt et al., 2017; Lau & Murnighan, 2005; Lindell et al., 2009; Lindell & Perry, 2000, 2012; Lion et al. 2002; Lopes, 1992; Marti et al., 2018; McBride et al., 2019; McClure et al., 2015; McLennan et al., 2014; Mileti & O'Brien, 1993; Morrissey & Reser, 2003; Paton & McClure, 2013; Paton & Buergelt, 2019; Reininger et al., 2013; Siegrist & Cvetkovich, 2000; Slovic et al., 1982; Slovic et al., 2002; Spittal et al., 2005; Terpstra & Lindell, 2013; Vinnell et al., 2020; Ward, 2021; Witte, 1992.

pation were evident in changes to the social and cognitive correlates of earthquake preparedness.

Compared to their non-participant counterparts, drill participants became less fatalistic, developed more realistic estimates of a serious earthquake occurring, and perceived preparing for earthquakes as less inconvenient and less difficult. Participants were also more likely to report that preparing for earthquakes could reduce damage to their home, improve their everyday living conditions, and enhance their capacity to cope with the disruptions created by earthquake consequences.

Vinnell et al. (2020) proposed that drill participation could be enhanced by adding a goal-setting component that encourages participants to both learn about possible actions that can be adopted and commit to completing one action following drill participation (McClure et al., 2015). Increased preparedness can also be enhanced by including a community engagement component to support learning and adoption (Adams et al., 2017).

Adams et al. (2017) reported that including community-based events (e.g., disaster planning meetings) or earthquake preparedness games increased people's scores on relevant social-cognitive indicators (e.g., self/outcome efficacy, personal responsibility) and their levels of preparedness knowledge compared to those who participated in the drill component only. The process and content issues discussed by Vinnell et al. (2020) and Adams et al. (2017) are summarized as an evaluation framework in Figure 7.1.

Further work, according to Vinnell et al. (2020), could consider extending the scope of the drill to encompass actions to take when outside (e.g., moving to a clear, open space away from potential hazards when walking outside, what to do when driving), including pre- and post-event longitudinal research, and assessing whether pre-event levels of preparedness influence decisions to participate. Vinnell and colleagues also called for attention to be directed to including theory-based evaluation. An example of the latter is discussed next.

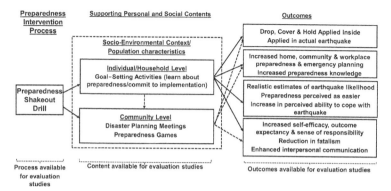

Source: Adapted from Adams et al., 2017; McClure et al., 2015; Vinnell et al., 2020.

Figure 7.1   The Shakeout Drill presented as an evaluation model

7.5.2   Theory-based evaluation

While the preparedness theories discussed in Chapter 3 are extensively researched, less attention has been paid to the origins of the scores, and reasons for the variability of these scores, in the research. It is argued here that the latter has implications for content and outcome evaluation.

When they participate in a survey, participants bring to this exercise their life experience in dealing with opportunities and challenges in home, social network, neighbourhood, community and work settings (e.g., advocating and planning for new community amenities, planning home renovations, working on marketing projects). These experiences foster the development of capabilities and relationships that have DRR implications.

For example, as people gain experience in successfully responding to life's challenges in personal, family, social or work-related settings, their progressive development of *self-efficacy* and *coping* competencies will be reflected in their scores on relevant questions in preparedness surveys. Similarly, accumulated experience in cooperating and collaborating with others to define and resolve issues in family, neighbourhood, social network and community settings will be captured in people's scores on variables such as *community participation*, *collective efficacy*, *sense of community*, *social capital* and *social support*. Hence, scores on variables implicated in how people socially construct their understanding of their risk, the strategies they develop to manage their risk (Lion et al., 2002; Paton et al., 2005; Paton, 2008) are developed from people's life experiences. Similar processes apply to people's experience with societal-level agencies.

Over time, people accumulate varying levels of direct and indirect (e.g., via media reporting) experience with scientific and risk management agencies. The quality of these experiences translates into people's scores on variables such as *empowerment*, *trust* and *linking social capital*. Taken together, this means that the scores recorded in surveys or the views expressed in interviews derive, wholly or partly, from people's life experiences and the knowledge, capacities, capabilities and relationships developed over their lifetime.

From this perspective, differences in the type and quality of people's accumulated experiences will be reflected in variance in their predictor variable scores (e.g., *collective efficacy*, *empowerment*, *trust*). If a theory incorporating these variables can account for significant levels of variance in preparedness, the theory can be used as a process evaluation tool, and its constituent variables can be subject to content evaluation. Reframing theories in this way offers another approach to intervention design and evaluation. An example of what this means is discussed by illustrating how the CET was used to design a community development-based DRR strategy.

Research into wildfire preparedness using the CET (Frandsen et al., 2012) informed the development of a wildfire preparedness program, the Bushfire-Ready Neighbourhoods (BRN) strategy. An evaluation study (Paton, Kerstholt & Skinner, 2017; Skinner, 2016) compared six BRN communities (whose members engaged with fire authorities to collaboratively develop locally relevant DRR activities) with six control (matched) communities that did not receive the BRN program. The pre-BRN assessment was conducted in 2014 and a comparative evaluation was undertaken in 2016.

The BRN strategy was built on a foundation of compiling community profiles and using this to plan activities specifically tailored to the circumstances and DRR goals in each community (Figure 7.2). The BRN interventions included activities to develop positive outcome expectancy beliefs, encourage community participation, and facilitate collective efficacy and other CET variables (Skinner, 2016).

Regarding people's outcome expectancy beliefs, stories from members of comparable communities (whose members had wildfire event experience) describing how preparedness enabled their response to fire events were incorporated as BRN development strategies (Figure 7.2). These activities enabled BRN community members to learn what worked from people living and working in circumstances similar to their own.

Interventions to develop community participation, collective efficacy and sense of community involved activities and exercises that identified and then developed community strengths. These activities were supported using community forums to develop and implement collective actions deemed by residents as being responsive to local needs (Figure 7.2). The development of positive outcome expectancy, community participation, and collective efficacy and empowerment were further supported using property fire safety assessments and wildfire survival planning workshops (Frandsen et al., 2012; Skinner, 2016). The relationship between CET variables, community engagement activities and DRR outcomes is summarized in Figure 7.2.

The 2016 evaluation found improvements in household preparedness. On average, in 2016, BRN-community members completed five more preparedness activities compared with 2014 baseline data (Skinner, 2016). Items where significant change occurred are summarized in Figure 7.2.

The evaluation revealed that participants in the BRN program, compared to those in the non-BRN control group, developed more comprehensive household wildfire response plans, took responsibility for, and developed a sense of personal ownership over, their household and community preparedness, and undertook significant structural preparedness activities (Skinner, 2016). In contrast, levels of preparedness in the non-BRN group remained largely unchanged from the 2014 baseline (Figure 7.2).

This evaluation study illustrates how theory can guide preparedness intervention planning by integrating risk management and community development approaches. Future work is needed to determine if such approaches are effective with less frequent and more unpredictable (e.g.,

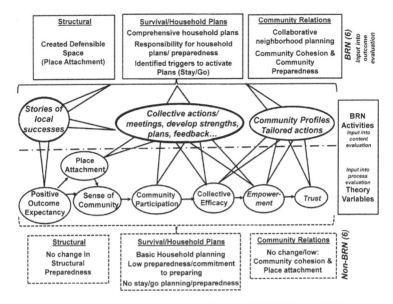

Source: Adapted from Paton, Kerstholt & Skinner, 2017; Skinner, 2016.

Figure 7.2 The relationship between the CET, community engagement strategies, and changes in preparedness in BRN and control communities

earthquakes) hazards and whether this model can operate in more socially complex urban settings, and whether the observed preparedness and relationship benefits are sustained over time, particularly in a context of changing community needs and membership. The two evaluation studies discussed focused on evaluating the effectiveness of pre-event DRR strategies. It is also important to consider how interventions delivered in recovery settings can be effective. How this might be accomplished is discussed next.

7.6 Evaluation in recovery and rebuilding settings

Researching outcome evaluation in recovery and rebuilding settings is complicated by it needing to accommodate diverse personal, social, economic and environmental outcomes (Scheinert & Comfort, 2014).

The specific outcomes experienced will differ from person to person, event to event, location to location, and from one country to another. It is, however, possible to adopt a generic approach that can encompass the core outcomes required to evaluate recovery and rebuilding outcomes.

Because it encompasses physical, psychological, social relations, and environmental outcomes, Quality of Life (e.g., using the WHOQOL-BREF; see https://www.who.int/tools/whoqol/whoqol-bref) represents an appropriate outcome evaluation measure in post-disaster settings (Papanikolaou et al., 2012; Paton & James, 2016; Paton et al., 2016; Sullivan & Sagala, 2020; Valenti et al., 2013; Walker-Springett et al., 2017). At present, there is no theory available to predict QoL outcomes. To compensate for this, a tentative process framework model is proposed. This, along with the content variables derived from the work of the above authors, is described in Figure 7.3.

Regarding the variables that could be subjected to content evaluation, Sullivan and Sagala (2020) identified roles for subjective social status, social identity, sense of shared fate, identification with the community, trust in authorities, the media and in community leaders, and type of displacement (in situ, temporary displacement, permanent relocation) as predictors of QoL. Paton et al. (2016) discussed how governance processes contributed to QoL by providing livelihood and lifestyle (e.g., skills training, business planning, reestablishing agricultural livelihoods) development opportunities that supported psychological, social relations, and environmental QoL outcomes. Paton and James (2016) identified how strengths-based interventions from government and NGO agencies contributed to well-being in recovery settings. A final process issue derives from Papanikolaou et al.'s (2012) comparative study of pre- and post-event scores across all QoL domains. Their finding that QoL domains recover at different rates, with environmental QoL being the last to emerge, calls for research to delve deeper into QoL process, content and outcome procedures. To support the latter, a tentative composite framework is depicted in Figure 7.3.

The model in Figure 7.3 depicts people's disaster-related relocation experience as the starting point. The relationship between relocation and QoL is mediated by social context resources, trust in authorities, community leaders and the media, empowerment, and post-displacement social identity and recovery status. Governance is depicted as playing a moderating

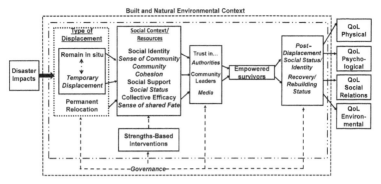

Source: Adapted from Papanikolaou et al., 2012; Paton & James, 2016; Paton et al., 2016; Sullivan & Sagala, 2020.

Figure 7.3   A tentative process, content and outcome framework applying QoL to evaluating recovery and rebuilding

role in the process, including how it facilitates adopting strengths-based approaches to community capacity development. This model should be regarded as tentative until subjected to systematic testing and evaluation.

7.7 Conclusion

The discussion of recovery and rebuilding experiences that concluded this chapter echoes comments made in Chapter 6 regarding how people affected by disaster, even when ill-prepared, actively seek ways to cope with and adapt to their circumstances. This outcome bodes well for finding support for the SFDRR Priority 4 call for adding a BBB or capacity development process to recovery and rebuilding initiatives. The next chapter builds on these observations to explore whether people can extend their recovery experiences by adding a learning dimension to it in ways that result in their transforming their experiences into enduring capacities and capabilities.

8 Transformative learning, capacity development and building back better

> On the occasion of every accident that befalls you, remember to turn to yourself and inquire what power you have to turn it to use.
>
> Epictetus (60–120 CE)

8.1 Learning from disasters

A significant challenge to effective DRR is people's reticence to prepare in ways that proactively reduce their risk and enhance their capacity to respond when disaster strikes (see Chapter 4). Given that societies and citizens the world over will experience progressively more frequent, intense and enduring environmental hazard events over the coming decades, particularly from climate change processes (e.g., flooding, hurricanes), this lack of preparedness is problematic. However, and ironically, the experience of disaster could act as a catalyst for developing sustainable DRR capacities and capabilities. The key lies, as identified by Epictetus, in whether societies and citizens wish to capitalize on the accidents that befall them, and whether they choose to inquire what power they have to turn it to use. This chapter focuses on inquiring into the power we have to turn disaster experience to use.

A significant contributor to people's preparatory reticence is their limited experience of the personal, social and societal problems environmental hazard events create. The consequent lack of tangible insights into what they could experience, and hence what they need to prepare for or recover from, can impede people's ability to anticipate their future DRR needs. When disaster strikes, however, these circumstances change.

The potential for disaster experience to motivate DRR capacity development derives from it offering societies and citizens in situ opportunities to acquire direct experience of the nature, intensity and duration of disasters and the hazard consequences they need to prepare for and be able to respond to. Disaster experience thus creates opportunities for reflection on the value of DRR. The question is whether this translates into citizen and societal capacity development.

This chapter discusses how while post-disaster capacity development is a reality, it cannot be regarded as a fait accompli of disaster experience per se. It will discuss how, when capacity development occurs, it does so in different ways and to varying extents, and sometimes it has no effect on capacity development.

8.2   Experience and inaction

Disaster experience can leave levels of preparedness unchanged (Comstock & Mallonee, 2005; Huang et al., 2016) and even result in people reducing their levels of preparedness (Johnston et al., 1999). One factor implicated in explaining such outcomes derives from the personal significance of people's experiences.

If an event does not create personally significant losses, preparedness behaviours can remain unchanged (Siegrist & Gutscher, 2008; Zaksek & Arvai, 2004) and even result in people discounting their future risk and negating their perceived need for additional preparedness (Johnston et al., 1999; Mileti & O'Brien, 1993). For example, Johnston et al. (1999) compared people's risk beliefs and preparedness levels before and after their experiencing low levels of ashfall hazards from the 1995 eruption of Ruapehu volcano (New Zealand). Johnston and colleagues found that while people's perception of volcanic risk increased, this was accompanied by their simultaneously reducing their level of preparedness; they discounted their need for future action. If, however, disaster experience activates people's psychological sense of uncertainty and insecurity, it can stimulate change in DRR beliefs and actions (Siegrist & Gutscher, 2008; Zakzek & Arvai, 2004).

Emotionally and psychologically demanding disaster experiences can challenge people's unrealistic optimism and prompt their interest in preparing but only in relation to the trigger event (e.g., an earthquake), and not to other relevant sources of risk in their environment; their interest in preparedness can thus be short-lived (Helweg-Larsen, 1999; Johnston et al., 1999). For instance, Johnston et al. (1999) found that experiencing low-level volcanic hazard consequences enhanced the salience of volcanic events in people's thinking but simultaneously reduced their interest in the more frequently occurring earthquake hazards they could experience.

The above findings do not, however, offer insights into whether the kinds of comprehensive and enduring shifts in capacity development envisaged by the SFDRR (UNDRR, 2016) can develop. To explore this issue, this chapter draws on survivors' accounts of their (major) disaster experiences to introduce three types of capacity development outcomes; *repurposed*, *emergent*, but relatively short-lived, change, and *transformative* learning. These processes are introduced next.

8.3 Capacity development: repurposing, emergent and transformative processes

The first process, *repurposing*, involves disaster survivors applying pre-existing personal and collective strengths (skills, experience, relationships, etc.) to the novel demands encountered in disaster recovery settings. For example, community organizations with experience of dealing with issues of local relevance (e.g., ratepayer associations) can apply their existing capacities and modes of working to resolving novel local recovery (e.g., securing resources) challenges (Paton et al., 2014; Williams et al., 2021). A second category can be described as *emergent* learning.

Emergent learning involves people developing *new* personal and collective and collaborative capacities in order to respond to novel recovery challenges. This can entail developing new neighbourhood groups and organizing their skills, experiences and connections to, for example, organizing house repair crews or developing social support resources for those in a neighbourhood (Paton et al., 2014; Williams et al., 2021).

Emergent groups may dissipate as recovery progresses and people return to normal functioning. Sometimes, however, they adopt a more enduring identity. This third category, *transformative learning*, results in creating new socio-structural institutions that become established within the fabric of social and societal life in ways that contribute to future hazard response capacity of the kind sought by the SFDRR (UNDRR, 2016).

Next, this chapter introduces repurposing, emergent and transformative learning outcomes, their origins, and the conditions required to enable progression from repurposing to emergent to transformative learning outcomes. Examples of *repurposing* and *emergent* capacity development are discussed first, and *transformative* processes are covered later in the chapter.

8.3.1 People's lived disaster experience: repurposing and emergent outcomes

Examples of *repurposing* and *emergent* outcomes are illustrated here by drawing on analyses of people's accounts of their disaster experiences from events in New Zealand, Australia (Tasmania and New South Wales (NSW)) and Taiwan (Irons & Paton, 2017; Paton, Jang & Irons, 2015; Paton et al., 2014; Williams et al., 2021). Across these events, people described repurposing as arising from, for example, their applying their pre-existing trade, DIY, building, or organizational skills to novel response tasks (e.g., assisting neighbours with home repairs, organizing work groups). In New Zealand and NSW, people described how their feelings of attachment to the city and to their suburb (cf., place attachment, city identity, sense of social responsibility) was repurposed into motivating their interest in engaging in collective recovery tasks. Emergent capacity development was also evident in each location.

In New Zealand, an example of emergent capacity development concerned the development of the family negotiation processes required to address issues ranging from deciding to remain in Christchurch during aftershocks to how the family could support local recovery activities to developing their stress management practices (e.g., managing aftershock experiences, relocation) (Paton et al., 2014). Emergent outcomes also manifest in the in situ development of neighbourhood self-help groups.

Emergent community groups arose to address urgent response needs (e.g., organizing repair crews, supporting elderly neighbours) (Paton et al., 2014). People's engagement in these groups helped develop their collective capacities (cf., collective efficacy), their sense of shared fate, and the social support resources used to sustain people's well-being and their commitment to act during numerous aftershocks. The effectiveness of these emergent neighbourhood groups was reinforced by emergent community leadership (Paton et al., 2014; McAllan et al., 2011).

Emergent local leaders played pivotal roles in empowering community action in two ways. One concerned encouraging social inclusivity (cf., sense of community) and actively involving as many people as possible in recovery tasks. The other derived from their representing local needs to government agencies and businesses in ways that helped secure the resources and information needed to support local recovery initiatives (Paton et al., 2014). These findings were echoed in the analysis of recovery experiences in Taiwan.

In Taiwan (Paton et al., 2015), the repurposing and emergent outcomes found echoed those in New Zealand (e.g., community self-reliance, local community leaders), albeit with some differences. For example, in Taiwan, local leadership was described as a repurposed resource (it was an emergent resource in New Zealand). This reflects the role local community leaders play in everyday community life in Taiwan (Jang & LaMendola, 2006).

Taiwanese survivors described their spiritual beliefs as being a repurposed resource. Pre-existing spiritual beliefs reinforced people's sense of purpose and perseverance in recovery tasks (cf., self- and collective efficacy), reinforced reciprocal support processes and social responsibility, helped manage stress, and strengthened their focus on securing positive outcomes for their future (Jang & LaMendola, 2006). Taiwanese respondents also identified *accepting disasters as part of life experience* and *co-existing harmoniously with nature* as repurposed adaptive capacities (see below).

The Taiwan analysis described the role of (devolved) governance practices in establishing local reconstruction centres to facilitate local recovery and community development as an emergent outcome (Paton et al., 2016). Some additional insights into emergent processes arose from analyses of response and recovery from wildfire events in Australia in 2013 and in 2020 (Irons & Paton, 2017; Williams et al., 2021).

Regarding the 2013 Tasmanian wildfire, Irons and Paton (2017) described how developing a Facebook page specifically to support recovery for this event created a unique emergent social identity linked to people's ability to share their personal and collective experiences over time. Secondary benefits included the Facebook page acting as a medium in which the emergence of people's sense of place attachment and sense of community functioned to support people's engagement in locally meaningful self-help activities.

Other emergent outcomes included developing information management processes that integrated top-down (e.g., agency, NGO) and bottom-up (community-based) communication in ways that empowered community recovery by increasing the timeliness, relevance and local meaningfulness of the information available to support meeting the needs of diverse survivor groups (e.g., elderly residents versus those with young children versus those evacuating) (Irons & Paton, 2017). These resources interacted to enable the emergence of a *collective intelligence* resource that supported local recovery planning and interventions (Irons & Paton, 2017; Arneson et al., 2017). In Tasmania, response activities were expedited by the availability of online, emergent community leaders who enabled the coordination of collective solutions to novel problems (Irons & Paton, 2017; Spialek & Houston, 2019). The final example of repurposing and emergent capacity development comes from a wildfire-affected community in NSW in 2020 (Williams et al., 2021).

In NSW, an example of repurposing involved how the local community co-operative, which had an existing structure and functions, volunteer network and prior experience serving the community, repurposed itself into a resource distribution centre for local farmers (e.g., distributing feed, fuel) (Williams et al., 2021). Williams and colleagues discussed how community events (e.g., the Light Up Cobargo festival, Cobargo Folk Festival) that took place in the main street and the showgrounds helped sustain people's sense of city identity (to the town and its built characteristics), with this spilling over into motivating people's involvement in city-based recovery activities.

Involvement in these place-based activities helped fuel people's sense of local (city) connectedness, with this underpinning the emergent development of, for example, social support resources, people's sense of gratitude for volunteer activities, acceptance of diversity, and the development of a sense of collective vision for the town's future. Williams and colleagues (2021) also shared how emergent change occurred even amongst those not directly involved in recovery activities. For example, some residents reported that witnessing the generosity and altruism in emergent groups fostered their experiencing gratitude for the volunteers involved in recovery and in regenerating city functioning (see also Chapter 6). The feelings of gratitude fuelled their developing a new sense of belonging to the city and its residents.

Local leadership was central to creating and maintaining local recovery initiatives based on activities in the showgrounds (Williams et al., 2021). For example, local leadership was implicated in how the volunteer groups that emerged in the showgrounds were subsequently institutionalized into the Bushfire Relief Centre that continues to operate and so represents a potential transformative outcome.

As in the other cases, key leader competencies included their being trusted, their ability to empower community participation, and their capacity to mobilize members' pre-existing skills (e.g., events management, administration), and direct them to managing local needs (e.g., sourcing generators, stock feed, organizing local activities). The leader role in establishing liaison mechanisms with government agencies and businesses was reiterated, as was their being crucial to ensuring that external agencies played complementary roles in community regeneration. However, Williams and colleagues highlighted how the effectiveness of leader roles was a function of the extent to which they are supported through policy and governance practices, with a lack of government policy continuity being implicated in emergent groups being sidelined and lacking the support needed to ensure their continuity. These observations echo those in Christchurch.

Christchurch survivors reported that discrepancies between agency functions and community needs contributed to residents feeling marginalized and disempowered (Paton et al., 2014). This was attributed to, for example, insufficient agency staff training and coordination mechanisms, lack of agency knowledge of local circumstances (Paton et al., 2014), and inadequate attention to recovery governance processes (Mamula-Seadon & McLean, 2015). Activities that disempower people diminish both people's trust in government agencies and their willingness to collaborate with them in recovery settings, so depriving communities of access to valuable resources (Monteil et al., 2020; Paton et al., 2014).

One approach to countering this problem involves using strengths-based interventions such as the *Linking Relief, Rehabilitation and Development* concept (Mosel & Levine, 2014) to support intervention planning and intervention. The intention is for government agencies and NGOs to support recovery by co-creating community capability, including through developing social capital by developing survivors' pre-existing historical, social, economic, spiritual and cultural strengths (Buergelt & Paton, 2014; Monteil et al., 2020; Mosel & Levine, 2014; Paton & James, 2016). This approach is not, however, without its challenges.

For example, Monteil et al. (2020) discussed how while bridging and linking social capital can enable the emergence of social cohesion amongst social network members, bonding social capital can impede recovery. This prompted their calling for more research into how social capital contributes to capacity development.

A possible explanation for Monteil et al.'s (2020) finding derives from how risk compensation bias (see Chapter 3) can stimulate the transfer of responsibility for recovery outcomes from citizens to government agencies (Paton et al., 2014). Another possibility derives from Seebauer and Babcicky's (2017) observation that trust in local government (linked to social capital processes) can increase people's reliance on government support in recovery settings and lessens their perceived need to take responsibility for managing their own recovery outcomes.

Consequently, effectively utilizing bonding social capital requires that recovery strategies explicitly call for citizen adoption of shared responsibility principles and acknowledge their responsibility for making complementary contributions to *their* recovery (Seebauer & Babcicky, 2017). This issue restates calls for exploring how governance processes are developed and enacted in socially complex, diverse recovery settings (Mamula-Seadon & McLean, 2015; Munene et al., 2018).

The above discussion illustrates how people's critical reflection (an important activity – see the Confucian epigraph in Chapter 6) on their disaster experience can stimulate repurposing and emergent learning. However, the repurposing and emergent changes described above did not always persist beyond the point where affected communities regained social stability following adequate levels of societal organization and functioning re-emerged. Consequently, if the kind of sustained capacity development envisaged by the SFDRR (Priority 4) is to occur, it becomes important to look beyond these examples to explore cases in which transformative learning occurred. This is explored next.

8.4 Transformative learning

This section explores whether disaster experience can generate fundamental personal and collective shifts in how societies and citizens perceive themselves and the world they inhabit, their socio-cultural-environment (built and natural) interdependencies, and the DRR beliefs, relationships and actions they develop. That is, as a transformational learning process that encompasses the complementary contributions of people, community, society and environment to the creation and maintenance of new approaches to DRR (Mezirow, 2008; Paton & Buergelt, 2019; Pelling, 2011). This topic is discussed in this chapter and culminates in offering a tentative model of transformative DRR learning.

8.4.1 Transformative DRR capacity development

This section revisits the Taiwan, New Zealand and Tasmania cases discussed above and compares them with analyses from Japan, Indonesia, Taiwan (Ho-Ping) and NSW to advance understanding of how transformative capacity development can occur. It commences with an example from Japan.

The 1914 Taisho eruption of Sakurajima volcano served as a catalyst for DRR transformation for the citizens of Kagoshima (Kitagawa, 2015). The process was triggered by the city mayor reflecting on how a major contribution to the loss of life and destruction the 1914 eruption visited on Kagoshima's residents was the inappropriate preferencing of scientific (for people to remain in situ) predictions over local people's knowledge (local people read local warning signs and argued for evacuation). Indeed, many survivors were those local people who disregarded government instruction and self-evacuated; they trusted their judgement and then took responsibility to evacuate themselves. The mayor's critical reflections on these issues informed the development of a transformative learning process that culminated in fundamental and enduring shifts in local DRR beliefs and practices in Kagoshima.

The transformative learning strategy instigated by the local mayor and a prominent seismologist applied community development principles to (re)building trust in civic authorities through a process of engaging and empowering citizen DRR (Kitagawa, 2015). This focused on enabling citizen capacity to (a) take and exercise responsibility for their own safety (personal *agency*), (b) engage in collective activities to develop their being *knowledgeable* about volcanic hazards and how to respond to them, and (c) appreciating that learning to co-exist with an environment in which volcanic hazards are an enduring fact of life is essential for effective DRR.

An important element of this strategy was it shifting from telling people what to do (a characteristic of the top-down communication approach previously used in Kagoshima) to developing people's ability to take responsibility for ensuring their safety (Kitagawa, 2015). Empowering people's DRR thinking in this way enabled them to understand *their* hazardous circumstances, the need to develop *their* knowledge of what *they* needed to do, and it created a social context conducive to sustaining these lessons that has persisted for over 100 years.

This transformative process was institutionalized through the development of *kyozon* (living with an active volcano – a social construct that reconciles the benefits and costs arising from living adjacent to a highly active volcano) and *kyojo* (helping each other through cooperative commitment) constructs (Kitagawa, 2015). These constructs advanced DRR capacities by enabling the collective adoption of cooperative and reciprocal practices and reinforced the value of people sharing responsibility for DRR. Another example of transformative learning occurred on the island of Simeulue (Indonesia).

Following a devastating tsunami in 1907, Simeulue islanders' reflection on their experience instigated the development of a new way of co-existing with the threat of tsunamis (Sutton, Paton, Buergelt,

Meilianda et al., 2020; Sutton, Paton, Buergelt, Sagala et al., 2020). The outcome was *smong*, a concept unique to Simeulue. *Smong* embodies people's knowledge of natural tsunami precursors, the urgent need to take personal responsibility to act if these precursors present, and the actions required to safeguard oneself from tsunami events (Sutton, Paton, Buergelt, Meilianda et al., 2020).

The transformative nature of the *smong* concept is apparent in its constituent knowledge and practices being actively passed through the generations by community elders passing on stories and song for over a hundred years. Its value was realized in the 2004 Indian Ocean tsunami. The low death toll on Simeulue, compared with that on nearby islands, resulted from *smong* triggering a rapid, effective and collective response from the islanders (Sutton, Paton, Buergelt, Sagala et al., 2020). Another example of transformative learning comes from Taiwan.

After their township was isolated by infrastructure damage following the 921 (or Chi Chi) earthquake (in 1999), the residents of Ho-Ping had to develop strategies to support their recovery. Transformative learning here culminated in a new approach to managing recovery labelled *community consciousness* (see Figure 5.3). This encompassed the development of new functional social relationships amongst community members, more empowering relationships with civic authorities, and stronger social–environmental relationships (Paton et al., 2016).

Finally, in NSW, Williams et al. (2021) identified how volunteer groups that emerged to manage recovery demands were subsequently transformed into the Bushfire Relief Centre. This remained in operation in 2021 and so represents a potential transformative outcome. Here, the key drivers of this transformation were social cohesion, place attachment, city identity and the role of local leadership in institutionalizing the Bushfire Relief Centre (Williams et al., 2021).

Taken together, the cases discussed above illustrate that while some disasters result in relatively short-lived emergent capacities, others culminate in more enduring transformative outcomes. The next section seeks to articulate the factors that facilitate the transition from emergent to transformative learning outcomes.

8.5 Modelling transformative DRR learning

In this section, cases where emergent outcomes occurred (e.g., Taiwan, Christchurch, Tasmania) are compared with those in which transformative processes emerged (Japan, Indonesia, Taiwan (Ho-Ping), NSW). This comparative approach is used to construct a model of transformative DRR learning (Figure 8.1).

Several factors – community cohesion, place attachment, social inclusivity, and the creation of empowering relationships between key stakeholders (community group members, NGOs, businesses, government agencies) – were present in all cases. These processes map onto, for instance, the community participation, social justice, collective efficacy, empowerment, trust, and sense of community constructs implicated (see Chapter 4) in facilitating coping and adaption (e.g., shared sense making capability, collaborative problem-solving, planning competencies) to novel, challenging disaster recovery circumstances (Earle, 2004; McAllan et al., 2011; Monteil et al., 2020; Silver & Grek-Martin, 2015). This suggests that these adaptive capacities (see Chapter 4) play roles in enabling the establishment of new social structural DRR processes, but do not appear, in themselves, to inform how these become established within the fabric of community life. Consequently, while they merit inclusion in a model of transformative learning (Figure 8.1), a question remains. What additional elements were present in the Kagoshima, Simeulue, Ho-Ping and NSW cases that stimulated the development of transformative learning outcomes?

Answering this question requires broadening the search for factors that could account for differences in outcomes between the cases. The first issue examined involves considering the relationship between local leadership and the context in which leadership is enacted.

The finding that local leadership was an emergent outcome in all cases precludes it representing a driver of transformative learning per se. However, other circumstantial aspects of leader experience can throw light on how local leaders can function as transformative learning agents over time.

Several circumstantial factors can adversely affect leader tenure and hence leader capacity to facilitate transformative learning. Factors such as emergent local leaders having to terminate their roles early (e.g., from exhaustion, family and livelihood requirements) and lack of leader succession planning have been implicated in this regard (McAllan et al., 2011; Thaler & Seebauer, 2019). Given that transformational learning takes time, any factor that limits leader engagement over time will reduce their ability to enable the conversion of emergent outcomes into enduring transformative learning outcomes. Support for this view can be gleaned from the case studies.

For example, in Kagoshima, the mayor's leadership tenure preceded and extended beyond the event itself, as did the roles of community leaders in Simeulue, Ho-Ping and NSW (Kitagawa, 2015; Paton et al., 2015, 2016; Sutton, Paton, Buergelt, Meilianda et al., 2020; Williams et al., 2021). Their consequent ability to maintain their roles over time ensured their availability to facilitate the development of enduring transformative social structural outcomes (e.g., *kyojo, smong*). In New Zealand and Tasmania, in contrast, the tenure of community leaders terminated when recovery stabilized, depriving communities of the leadership tenure necessary to consolidate emergent outcomes into transformative ones (Irons & Paton, 2017; McAllan et al., 2011; Paton et al., 2014).

While issues such as local leader selection and training remain salient issues (McAllan et al., 2011; Thaler & Seebauer, 2019), post-disaster capacity development strategies must ensure that leader tenure and succession planning issues (over potentially prolonged periods of time) are in place (McAllan et al., 2011). This discussion supports the inclusion of local leadership in a model of transformative DRR learning (Figure 8.1). Another line of inquiry worth pursuing concerns local leader engagement in *local* recovery governance development and implementation (McAllan et al., 2011; Munene et al., 2018).

8.6   Governance and transformative DRR learning

Countries that invest in governance policies and the institutional structures and relationships required to implement them enhance their capacity to create social and societal frameworks capable of enabling the development and deployment of DRR strategies at both national and local levels (UNDRR, 2020). In this context, exploring how local governance processes are developed and enacted, and by whom, can offer additional perspectives on both how transformative learning can arise and how it can be consolidated into enduring DRR capacities.

A significant reason for including local DRR governance (and its ability to complement its national counterpart) derives from the fact that large-scale disasters create geographically, socially and temporally diverse impacts. Consequently, the social and societal consequences that need to be managed will vary from place to place and over time. National governance processes, which typically take a more expansive, top-down approach, can be challenged by such circumstances (Mamula-Seadon & McLean, 2015; Munene et al., 2018). In contrast, local or devolved DRR governance processes are more responsive to local-level variability, making them better suited to facilitating how unique and varied local recovery and rebuilding needs can be met (Munene et al., 2018).

Local governance has been recognized as a significant driver of capacity development in response to social and environmental change and localized disaster impacts within local jurisdictions (Dhakal, 2012). Local governance processes are also better equipped to support shared responsibility perspectives in ways that enable citizens, government agencies, NGOs and businesses to play complementary roles in emergent and transformative DRR (Mamula-Seadon & McLean, 2015; McNamara & Buggy, 2017; Sarzynski, 2015; UNDRR, 2020).

It is worth noting that a more effective outcome will ensue when local and national DRR governance complement one another. Because it can, for example, facilitate distributed capability (i.e., enable adoption of DRR lessons in all jurisdictions, and not just in those affected by a specific event), expedite the strategic sourcing and distribution of resources needed at local levels, and support development through regulatory frameworks, national governance remains crucial. Hence, modelling how national and local governance can complement one another merits inclusion in a transformative learning model (Figure 8.1). It is proposed here that the effectiveness of local governance is enhanced by local leader involvement in its development and implementation (Figure 8.1).

In Kagoshima, Simeulue, Ho-Ping and NSW, local leaders were the local mayor or highly respected community elders and citizens with pre-existing responsibilities for managing city or community affairs before, during and after their respective disasters (Kitagawa, 2015; Paton et al., 2016; Williams et al., 2021). Hence, they occupied roles which encompassed their being responsible for developing and implementing local governance mechanisms that were responsive to local circumstances and needs, with this perspective reiterating the importance of including institutionalized local leader roles in a model of transformative learning (Figure 8.1).

In contrast, in the New Zealand and Tasmanian cases, emergent local leaders filled these roles only during the recovery phases of their respective events, lacked formal leader responsibilities, and often found themselves in conflict with formal leaders (Mamula-Seadon & McLean, 2015). These circumstances may have limited their opportunities for engaging in the governance processes present in the Kagoshima, Simeulue, Ho-Ping and NSW cases. Another informative characteristic of local governance in the Kagoshima, Simeulue and Ho-Ping cases was it emerging through bottom-up community engagement processes that empowered citizen involvement in consolidating community-led transformative learning outcomes (Kitagawa, 2015; Paton et al., 2016; Sutton, Paton, Buergelt, Sagala et al., 2020).

In Tasmania, in contrast, no local governance process emerged to facilitate the maintenance and consolidation of emergent social processes (Irons & Paton, 2017). In New Zealand, local governance processes were developed by the national government. However, while local representation was present, the predominantly top-down approach adopted lacked the flexibility to provide high levels of support for emergent local initiatives (Mamula-Seadon & McLean, 2015). Similar conclusions regarding the effects of a national–local governance disconnect have been documented in other studies.

Thaler and Seebauer (2019) found that top-down governance practices that limit or consign citizen involvement to performing support roles rather than engaging them in their own recovery diminishes the effectiveness of DRR governance. Thaler and Seebauer thus argued that including bottom-up, citizen-driven governance processes in DRR governance is important for capacity development. Local leaders can play important roles in this context.

Hence, local leader engagement in developing and applying local governance in ways that complement its national counterpart, merits inclusion in a model of transformative DRR learning (Figure 8.1). This view is consistent with findings from governance research calling for coordinated whole-of-society, multi-level approaches (Bulkeley & Betsill, 2005; Mamula-Seadon & McLean, 2015). This work reiterates the value of including links between leader roles and local governance in a model of transformative learning (Figure 8.1).

By enabling strong stakeholder engagement, multi-level approaches offer greater potential for reconciling the need for governance processes to facilitate both societal stability in everyday life with a capability to enable adaptive responses to the complex, dynamic and unpredictable social and environmental challenges that characterize disaster response and recovery settings (Beunen et al., 2017; Karpouzoglou et al., 2016; Munene et al., 2018). These authors expand on this idea to advocate for conceptualizing DRR governance as an iterative process capable of evolving and adapting to change over space and time. The governance development and application processes described in Kagoshima and Ho-Ping illustrate how the iterative, adaptive governance processes proposed by these authors could operate. Another facet of the processes operating in Kagoshima and Ho-Ping was their focus on developing adaptive capacities.

In Kagoshima and Ho-Ping, the transformative DRR learning outcomes implemented focused less on preparing for volcanic or earthquake hazards per se, and more on developing adaptive capacities capable of enhancing social responsiveness to any (future) challenging circumstance. For instance, in Kagoshima, the *kyojo* process that facilitates people's reciprocal and cooperative commitment to DRR initiatives can also support collaborative responses to any challenging situation, including those unconnected with DRR. These thus represent adaptive capacities that are available to help enact local governance policies irrespective of the circumstances in which they are applied (e.g., adapting to volcanic hazards versus climate change processes). The approach advocated here reiterates the important role citizen participation can play in supporting flexible and responsive multi-level governance (Uittenbroek et al., 2019).

The benefits of accommodating public participation in multi-level governance practices include its capacity to enable access to and utilization of local knowledges, values and expertise in emergent and transformative

DRR processes (Sarzynski, 2015; Uittenbroek et al., 2019). Expanding the range of knowledge, competencies and perspectives available to enable the design and implementation of DRR resources can be facilitated using adaptive governance processes that are more responsive to changes in hazardscapes (Beunen et al., 2017; Munene et al., 2018). Hence, including relationships between social relationship capacities (e.g., community participation, inclusivity, collective efficacy, social responsibility, trust) that enable transformative learning and the local leadership, local governance and national governance required to enact and sustain transformational outcomes merit inclusion in a model of transformative learning (Figure 8.1).

However, Uittenbroek et al. (2019) warn that such benefits will only be realized if strategies are implemented to safeguard against citizen participation being relegated to playing tokenistic roles in governance processes. Achieving this outcome is more likely if the development of local governance is complemented by applying community development strategies to ensure that it empowers all relevant stakeholders using participatory approaches that facilitate learning about and developing innovative solutions to manage evolving hazardscapes (Djalante et al., 2011; Karpouzoglou et al., 2016; Munene et al., 2018; Sarzynski, 2015; Uittenbroek et al., 2019).

In DRR contexts, however, it is important to remember that governance functions within several nested environmental contexts. Consequently, it is pertinent to consider how people's engagement in relevant environmental settings influence transformative outcomes. The environmental contexts discussed here include city, neighbourhood/location (place), and natural environmental settings. The first issue discussed is city identity.

8.7 Environmental context: city, place and natural settings

A body of evidence exists that attests to how the effectiveness of city (local) governance is influenced by people's sense of city identity (Healey, 2006; Kong, 2007; Peng et al., 2020; Williams et al., 2021). This suggests it would be beneficial to explore how the quality of people's engagement in city life influences how governance processes affect capacity development.

8.7.1 City identity

People's relationship with the city in which they live can influence their motivation to engage in DRR. For example, participation in city-based festivals linked to DRR and recovery processes can facilitate the development of trust in risk management agencies and help maintain people's interest in hazard preparedness and recovery interventions (Bhandari et al., 2010; Kitagawa, 2015; Williams et al., 2021). To these examples can be added evidence demonstrating how city identity acts as a medium for enabling the effectiveness of city (local) governance by facilitating place-based connections between people and events over time (Healey, 2006; Kong, 2007; Peng et al., 2020; Williams et al., 2021). So, how might city identity fulfil this motivational role?

Fundamentally, people's identification with, for example, a city's prominent visual characteristics (e.g., location and topographical characteristics, architectural elements), aesthetic parks, gardens, reserves and coastal scenery, and the socio-cultural activities it affords its citizens underpins how city identity motivates action (Adams et al., 2017; Bhandari et al., 2010; Kong, 2007; Williams et al., 2021). People's engagement with these diverse city characteristics (e.g., architecture, festivals) over time fosters the emergence of shared values, norms and patterns of behaviour that influence how people, individually and collectively, interpret and respond to environmental challenges and engage in emergent and transformative learning (Bhandari et al., 2010; Williams et al., 2021). This work warrants city identity being included in a model of transformative learning (Figure 8.1). However, the city is not the only environmental phenomenon within which capacity development could emerge.

Other candidates for environmental influences on transformation include place attachment and identity (De Dominicis et al., 2015; Frandsen et al., 2012; Paton, Buergelt & Prior, 2008; Silver & Grek-Martin, 2015). Discussion of the role of place attachment in transformative DRR commences by considering the relationship between city and place identities.

8.7.2 Place: locational and neighbourhood influences

Cities are environmentally complex. They comprise several locations, each characterized by their diverse relationships and interdependencies with various built and natural environmental (e.g., river, architectural features) settings (Rademacher, 2015). The specific location in which people live thus results in their being connected to place (e.g., neighbourhood, specific locality) and to city, with 'place' being an embedded spatial feature for those living in a city.

Place influences the hazards people experience (e.g., how close a person lives to a river), and this varies from one location/neighbourhood to another. Given that where people live (place) affects their risk, their perceived attachment to and identification with place can influence their DRR deliberations and the strength of their engagement in relevant DRR and recovery processes (Bhandari et al., 2010; Frandsen et al., 2012; Monteil et al., 2020; Seebauer & Babcicky, 2017; Silver & Grek-Martin, 2015; Williams et al., 2021).

The position advocated here is that city identity and place attachment/identity represent nested locational influences on people's DRR thinking, with place attachment and identity operating at the locality, home or neighbourhood level, and city identity representing an overarching environmental context in which place and their hazardscapes are embedded (Healey, 2006). This position has been echoed in studies by Bajek et al. (2008), Bhandari et al. (2010), Kitagawa (2015) and Williams et al. (2021). Consequently, modelling interdependent roles for city identity and place attachment/identity merit inclusion in a model of transformative learning (Figure 8.1). However, an issue that has been less extensively canvassed in studies of place and DRR relationships concerns how these relationships can be influenced by cultural characteristics (see Chapter 5). This issue is discussed next.

8.8 Cultural dimensions, place and DRR

The citizens of Kagoshima, Simeulue and Ho-Ping are representatives of collectivistic cultures. In general, for members of collectivistic cultures, relational and locational community coincide; people's relationships are embedded in the geographical locations in which they reside (Figure 8.1). This increases the likelihood that a driver of DRR, place attachment, overlaps with the geographical distribution of the hazard phenomena people must prepare for.

It can be postulated that this overlap increases the prospect of residents in a given area being motivated to engage in creating and enacting collective responses and solutions to shared local problems (Mamula-Seadon, 2018; Monteil et al., 2020; Seebauer & Babcicky, 2017; Silver & Grek-Martin, 2015). Hence, for the residents of Kagoshima, Simeulue and Ho-Ping, the synergy between locational identity (that existed pre- and post-disaster) and place could have contributed to inspiring their enduring commitment to creating locally meaningful DRR processes (e.g., *smong*, *kyozon*).

In contrast, the New Zealand and Tasmanian research demonstrating emergent, but not transformative, outcomes was situated in more culturally individualistic populations (Paton, Jang & Irons, 2015). This means that, prior to disaster occurring, people's relationships were predominantly described by their shared interests and social affiliations (as members of relational communities) rather than from their geographic (locational communities) locality per se.

The disasters in New Zealand and Tasmania thus disrupted people's ability to sustain their relational community ties and, instead, it became necessary for them to forge new relationships within their neighbourhoods (i.e., as members of emergent locational communities). This resulted in people developing emergent superordinate locational identities within the neighbourhoods in which their recovery activities occurred (Paton et al., 2014; Spialek & Houston, 2019). However, a similar argument cannot be applied to the NSW findings (Williams et al., 2021) so further consideration of this issue is warranted (see below).

In New Zealand and Tasmania, as recovery conditions stabilized, the motivational influence of the 'shared fate' social identity ties sourced from people's collective disaster experience diminished over time (Ntontis et al., 2020; Paton et al., 2015; Silver & Grek-Martin, 2015). As conditions stabilized and opportunities for normal social opportunities returned, residents' relational social identities re-emerged.

At the same time, the fading sense of neighbourhood-based place identity progressively reduced the sense of neighbourhood connectedness postulated as a motivator for capacity development in the collectivistic cultural examples in Japan, Indonesia and Taiwan. Hence, collectivistic cultural orientations can be invoked as a contextual influence on how place attachment can motivate transformative capacity development.

In contrast, in the more culturally individualistic New Zealand and Australian samples, the re-emergence of relational communities removed this influence on transformative learning.

This position argues for transformative DRR learning in more individualistic cultures to include strategies to maintain emergent superordinate locational (e.g., neighbourhood) identities and integrate them within prevailing relational social identities. The latter argument could offer insights into the development of transformative learning in the more culturally individualistic NSW setting (Williams et al., 2021).

Williams and colleagues (2021) discussed how social cohesion, place attachment, city identity and effective local leadership combined to facilitate the institutionalization of the Bushfire Relief Centre. This process could have created, and sustained, the kind of superordinate identity introduced as being important above. Hence, strategies that integrate local community-based leadership, governance and community development could offer an approach to creating the superordinate identity required to facilitate transformative learning in countries in which relational community membership prevails.

In addition to their socio-cultural similarities, the Japanese, Indonesian and Taiwanese examples of transformative learning had something else in common; they all involved socio-environmental co-existence process in their capacity development processes (see Chapter 5). This draws attention to the final environmental level of analysis that could be included in a transformative learning model, the natural environment and people's relationship with it.

8.9 Socio-environmental relationships and co-existence

Social–environmental relationships are central to DRR (Buergelt & Paton, 2014; Karpouzoglou et al., 2016; Twigg, 2015). Evidence exists which suggests that the more people hold social–environmental co-existence beliefs, the more likely it is that these relationships influence people's capacity development (Charlesworth & Okereke, 2010; Paton, Buergelt & Campbell, 2015; Woodgate & Redclift, 1998). The capacities developed in Kagoshima (e.g., *kyozon* and its emphasis on learning to co-exist with an active volcano), Simeulue (e.g., *smong* values understanding and being responsive to natural tsunami warning indicators) and Ho-Ping (environmental co-existence beliefs for living with hazards) all included social–environmental relationships. Comparable outcomes were not found in New Zealand or in the Australian cases. This points to the value of including environmental co-existence beliefs and practices in a model of transformative learning (Figure 8.1; see also Figures 5.2 and 6.2).

8.10 Pulling the transformative DRR learning threads together

This section pulls the several threads introduced above into a conceptual transformative DRR model. The conceptualization of transformative learning offered here depicts disaster experience as the catalyst for new ways of thinking that derive from people's active reflection on their experience (Figure 8.1). The relationship between this reflection and transformative outcomes is mediated by several social and societal processes that emerge from people's disaster experience.

These emergent mediating processes include several social and relationship competencies (e.g., community participation, collective efficacy, empowerment, trust) recognized for their role in facilitating adaptation to challenging circumstances (see Chapter 4). These core processes underpin emergent outcomes. These can be regarded as providing the social or community level of adaptive capacity required to provide a foundation for transformative learning. The model in Figure 8.1 postulates that the conversion of emergent to transformative learning is mediated by local leaders proactively empowering community action and developing functional links with external agencies, businesses and NGOs in ways that ensure the availability of the resources required to make transformational learning possible. Lesser roles extend to include their involvement in local governance development and implementation. The model also postulates that complementary contributions from local and national governance make important contributions to transformational learning, particularly when learning is embedded in city, place and environmental settings that motivate interest in developing and maintain transformative DRR outcomes. In addition, cultural orientation (collectivistic versus individualistic) is postulated to influence how place attachment and identity support the development of transformative learning outcomes. These relationships are summarized in Figure 8.1.

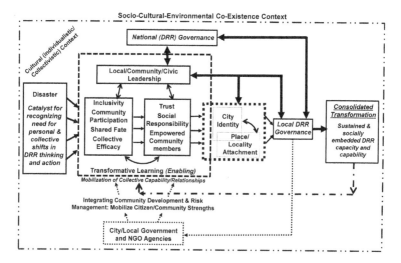

Figure 8.1   Summary of the DRR transformative learning process

8.11 Conclusion

This chapter conceptualized post-disaster capacity development as arising from repurposing, emergent and transformative learning processes. While each can enable coping and adapting to novel and challenging circumstances, only transformative learning ensures that learning outcomes become embedded in the socio-cultural-environmental fabric of societal, community and household life in ways that can support DRR processes into the future. This outcome reflects how transformative learning results from people making fundamental shifts in how they perceive themselves and their world, how they relate to each other, and how they think about and act towards environmental hazards (Buergelt et al., 2017; Mezirow, 2008). The transformative learning outcomes discussed here illustrate how the UNDRR (2015) goal of enabling enduring capacity development outcomes could be realized. The proposed model is summarized in Figure 8.1. Future research is needed to test this model across diverse settings.

9 Conclusions and future issues

9.1 Knowing DRR for the first time

> We shall not cease from exploration, and the end of all our exploring will be to arrive where we started ... and know the place for the first time. (T.S. Eliot, *Little Gidding*, 1942)

While activities such as risk analysis, preparedness and recovery have been central to the field of disaster management for decades, the priorities and principles (e.g., emphasizing all of society, multi-level approaches to sharing responsibility for DRR) developed by the SFDRR provided, to paraphrase Eliot's (opening quote) words ... a way to explore (disaster management) ... and know it for the first time. Additional contributions to the latter derived from the SFDRR including all-hazards, social diversity, social justice and cross-cultural perspectives in DRR and adding the BBB construct to how recovery planning and intervention is conceptualized.

Despite these achievements, it is, as Eliot states, vital that we shall not cease from exploration; in an era when humanity faces growing threats from evolving, uncertain and unpredictable hazards, developing ways to know DRR for the first time must become the norm. Some ideas for these explorations are discussed in this closing chapter. The first, concerning the implications of the lack of a conceptual basis for integrating SFDRR Priorities (Munene et al., 2018), is explored by considering whether adopting a socio-cultural-environmental framework can inform the development of a more cohesive framework.

9.2 A socio-cultural-environmental framework

This book introduced DRR by reiterating the importance of conceptualizing it in terms of socio-cultural-environmental interdependencies. It is proposed that adopting a socio-cultural-environmental conceptualization can inform the development of a cohesive DRR framework in one of two ways. One is by it affording opportunities to inform the development of socio-environmental co-existence beliefs that entail societies and citizens developing beliefs and practices that reconcile their ability to secure sustainable benefits from their environment with their developing the capacities to reduce their risk and enhance their capacity to respond to, recover from, and learn from, future hazardous experiences. A second is through it offering opportunities for adding a transdisciplinary focus to DRR. This is discussed in more detail below.

However, reconciling sustainable lifestyle and DRR beliefs and practices raises governance issues. If this reconciliation approach has merit, additional work will be required to explore its social and governance implications. Grounds for giving this some thought certainly exist. Projected increases in the environmental threats humankind will face over the coming decades highlight the importance of continuing to explore how these, often conflicting, outcomes can be sustainably achieved. Conflicts can arise from, for example, people's environmental beliefs and from differential mobilization of environmental beliefs (see Chapters 3 and 4).

For instance, people's willingness to embrace socio-environmental co-existence is influenced by how societies and their citizens conceptualize their socio-environmental relationships. If this relationship is defined in more anthropocentric terms, developing socio-environmental co-existence beliefs and behaviours becomes a more challenging endeavour (see Chapter 3). There is, however, evidence of general shifts towards more ecocentric ways of thinking (Buergelt et al., 2017). Notwithstanding, facilitating the kinds of whole-of-society shifts towards ecocentric, socio-environmental co-existence beliefs and practices is an area that merits inclusion in future research programs. The benefits of doing so were introduced in Chapter 2.

While scientific endeavours directed towards understanding disaster risk remain vital, Chapter 2 discussed how their contribution to DRR can only be realized by appreciating how personal (e.g., emotions), social (e.g., social-cognitive biases) and cultural (e.g., cultural interpretations of environmental threat – see Chapter 5) processes influence people's interpretation of these data and whether they use, or do not use, these data. Chapter 2 discussed how people can misinterpret scientific hazard behaviour data (e.g., return period), be unaware of the relevance of hazard behaviours (e.g., intensity, duration), or use behavioural data inappropriately (e.g., relying on warnings to trigger preparing). These examples illustrate why conceptualizing DRR in terms of socio-environmental interdependencies can be beneficial.

For example, the social side of this relationship will benefit from developing people's interpretive capabilities in ways that enable their being able to make effective use of scientific data. However, attention must also be directed to the environmental side of the relationship by developing the capacity of scientists and risk managers to communicate environmental risk data in ways that engage its recipients and enhance the meaningfulness of these data for their citizen recipients. That is, effective DRR requires that these activities are integrated (multi-level, shared responsibility).

Strategies to support the development of such cooperative and collaborative, shared responsibility relationships include facilitating opportunities for people to actively engage (e.g., citizen scientist groups, or NERTs) in discussing hazard consequences, their origins and implications, and using local hazard distribution maps and other interactive tools to provide people with more tangible insights into hazard data and its personal and local implications (McClure, 2017).

This approach, reconciling the needs of those who use environmental risk data and those who produce it, illustrates how a socio-environmental framework can support conceptualizing shared responsibility by illustrating how Priority 1 (understanding disaster risk) and Priority 4 (preparedness/BBB/capacity development), rather than being seen as discrete elements, can play more seamless and complementary roles in DRR. The latter draws attention to another area in which the development of governance processes could inform socio-environmental integration goals. Fully realizing the benefits of the relationship between risk data and preparedness processes will entail directing attention to other preparedness issues.

9.3 Future preparedness issues

Preparedness consists of several functional preparedness categories (see Table 4.1 and Figure 4.1). Depicting them individually, as they are in Figure 4.1, draws attention to another area for future exploration.

Typically, functional category items are combined into a single list (e.g., in public education brochures, websites) and represented as a single dependent variable in preparedness research. Treating functional preparedness categories as a homogenous set of activities raises several issues. One concerns how this approach increases the likelihood of reactivating the *single action bias* (see Chapter 3), resulting in people reverting to adopting single, low-cost preparation options and to discount the value of others, particularly the higher-cost but higher utility structural preparedness items (Finucane et al., 2000; McClure et al., 2009).

Another problem arises when combining preparedness items. It makes it more difficult to link risk data and preparedness functions. For example, intensity (see Chapter 2) data is most suited to targeting structural preparedness, whereas duration (e.g., aftershock sequence) data links more effectively to identifying long-term survival, and family and community capacity-building preparedness (Crozier et al., 2006; McClure, 2017). These issues introduce a need to ask whether preparedness functions should be separated rather than being combined.

In addition to the risk data-preparedness issue raised in the previous paragraph, other grounds exist, based on differences in the decision and task demands associated with each category, that warrant arguing that the functional preparedness categories should be seen as separate entities. For instance, adopting survival items (e.g., storing food/water) is a low-cost activity that imposes low demands on people's skills and time commitments (e.g., acquire and fill water containers). In contrast, livelihood preparedness requires analysing one's occupational circumstances and its exposure to risk, discussing preparedness and business continuity issues with employers, and planning how to respond to various contingencies (e.g., unable to get to work, being injured). Similarly, if they are to contribute to community response planning, people have to commit time to attend meetings, organizing meetings, reconciling their and others' needs and so on. These are demanding of people's time, negotiating and planning skills. Given the differences in the skill, competence and relationship demands associated with different functional categories, it can be postulated that each could have its own antecedents.

Qualified empirical support for this has emerged in studies of earthquake (Paton, Anderson et al., 2015) and wildfire (McLennan et al., 2014) preparedness. Future work is needed to systematically assess the respective antecedents of each functional preparedness category and, if distinct antecedents are identified, a predictor model will be required for each category. One component missing from preparedness theorizing is its links to hazard behaviour (see Chapter 2) data. Adding this dimension could facilitate providing information in ways that model relationships preparedness precursors, hazard behaviours and hazard consequences (Crozier et al., 2006; McClure, 2017). Another area preparedness research could explore concerns the sources of preparedness variable scores.

9.4 Community development and DRR

The majority of preparedness information is sourced from questionnaire data. However, questionnaire data offers few insights into the sources of people's scores or the origins of the differences that arise between people's scores. Chapter 7 introduced how people's scores on several preparedness theory (see Chapter 4) variables derive from people's lived experiences.

For example, experience in resolving life challenges will be reflected in people's *self-efficacy* scores, the quality of their social network and neighbourhood experiences in their *collective efficacy* and *social capital* scores, and their experience with government agencies will be reflected in their *empowerment* and *trust* scores. According to this position, scores on several key variables originate from people's life experience rather than from risk management interventions per se. Examples of variables that can derive from life experience and community development are summarized in Figure 9.1.

The fact that several key variables derive from life experiences supports calls to shift the core emphasis in DRR from hazard- and risk assessment-based methods to more socially oriented approaches that afford community development a more central role in contemporary

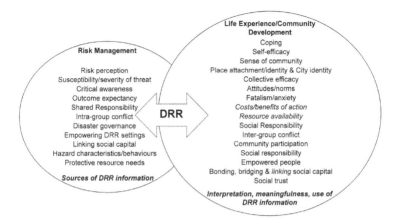

Figure 9.1 The respective contributions of life experience/ community development and risk management variables to DRR

DRR research and practice (Ismail-Zadeh et al., 2017). The content of Figure 9.1 introduces additional governance issues.

Given the preponderance of life experience variables that influence DRR outcomes, it becomes pertinent to inquire how disaster governance can accommodate the substantial influence of activities that occur outside the general remit of DRR. Grounds for considering this issue are discussed next.

Variables that tap into life experiences are amenable to change using community development-based approaches (e.g., providing structured personal and collaborative social opportunities to foster community participation, collective efficacy, empowerment, social capital, and trust). This approach assigns risk management (e.g., providing information on hazards, risk, and DRR options) strategies to playing complementary rather than leading roles in DRR processes (Ismail-Zadeh et al., 2017; Paton, 2017). Activities available to support the development and application of community development strategies are described under the Facilitating Sustained Preparedness heading in Table 7.4. If it can be developed, the role of governance in this context can have other benefits.

Important considerations here include how community development-based approaches can increase the cost-effectiveness of developing, implementing and maintaining local DRR processes. For example, community development approaches can profit from the benefits of local leadership (see Chapter 9) and provide opportunities for peer-based learning and mentoring (to support those developing their preparedness), and advocacy (to assist those disempowered and give them a voice in DRR processes) that represent cost-effective approaches to developing the cooperative and collaborative approaches to DRR called for by the SFDRR (see Chapter 7).

The adoption of community development-based approaches also enhances the potential for local groups (e.g., citizen scientist groups, NERTs, Community Boards, Jishubo) to play more direct roles in local DRR in ways that can expedite the inclusion of vertically integrated approaches to shared responsibility in DRR processes. The opportunities for devolved responsibilities so created are better suited to targeting local circumstances, making this a cost-effective option to pursue.

Furthermore, community development-based approaches can foster the emergence of capabilities that enable people's ability to deal with all life's challenges and opportunities (in locally relevant ways over time), and not in those circumstances in which people face demands from infrequent natural hazard events. This increases both the opportunities available for maintaining people's adaptive capacities and their availability to support DRR goals over the longer term.

Community development-based approaches are more consistent with the SFDRR calls for developing collaborative and cooperative approaches to DRR and for adaptive governance (see below). Exploring what is involved in putting these ideas into practice will require additional research, including through making greater use of transformative learning.

9.5 Transformative learning

A significant contribution of the SFDRR to DRR is through its inclusion of the BBB or capacity development construct. In societal contexts defined by generally low prevailing levels of DRR knowledge and actions

CONCLUSIONS AND FUTURE ISSUES   143

(see Chapters 2, 3 and 4), and in circumstances in which people's exposure to large-scale hazard events is ever-increasing, taking advantage of the learning and developmental opportunities afforded by people's disaster experiences represents a significant DRR resource. This was discussed in Chapter 8 along with the development of a tentative capacity development model (see Figure 8.1). This model can inform the development of research questions and hypotheses to test the validity and relevance of its predictions. The introduction of a transformative learning perspective raises other issues.

The transformative capacity development process described in Figure 8.1 illustrated how a catalyst for transformative learning, people's reflection on their tangible disaster experience, motivated individual and collective efforts to develop locally relevant and novel ways of thinking and action. These outcomes were then institutionalized in ways that ensured their making enduring contributions to social and societal DRR. Given that people the world over face growing exposure to unpredictable and dynamic hazardscapes, the need for transformative learning will become increasingly important over the coming decades.

However, even if citizens and societies engage in post-disaster capacity development, once transformative outcomes develop, they become the new norm. In a more unpredictable future, earlier transformative outcomes may, or may not, be capable of accommodating the risk implications emanating from the evolving hazardscapes people will face.

If a proactive approach to DRR is to be developed, it becomes pertinent to ask how transformative learning could be facilitated in pre-event contexts and in the absence of the kinds of disaster experiences described as essential catalysts for transformation in Chapter 8. What can be done to challenge people's existing beliefs and actions, encourage their anticipating future unpredictable and dynamic hazard- and risk-scapes, and enable multi-stakeholder engagement in collaborative and cooperative efforts to continuously and proactively evolve their DRR capacities and capabilities in the absence of actual disaster experience? Approaches that could be considered here include using multi-stakeholder planning techniques such as scenario planning and using exercises such as the Shakeout Drills introduced in Chapter 7 (Adams et al., 2017; Paton & Buergelt, 2019; Vinnell et al., 2020). More work on this vital topic is urgently required.

The model depicted in Figure 8.1 points to some other ways in which DRR might develop.

The transformative learning model described in Figure 8.1 draws attention to the need for representatives of different disciplines (e.g., physical and social scientists, geographers, governance and political scientists). This makes it pertinent to ask what can be done to ensure that the representatives of these diverse disciplines can cooperate and collaborate to facilitate transformative learning. One possibility is to adopt transdisciplinary strategies.

9.6 Transdisciplinary strategies

In circumstances where the coordinated activity of diverse stakeholders is required to ensure effective DRR outcomes, the adoption of transdisciplinary approaches to research and planning can be beneficial (Ismail-Zadeh et al., 2017). Transdisciplinary methods necessitate shifting the emphasis in DRR planning from risk assessment-based methods to those using co-production approaches. Co-production entails representatives from diverse disciplines, who do not ordinarily work together, and who frequently represent competing perspectives and interests, collaborating to integrate their combined knowledge, expertise and experience into a unified approach to confronting complex, multi-faceted DRR problems. This has implications for how the SFDRR Priorities are discussed.

For instance, their current presentation as discrete, numbered categories increases the likelihood of their stimulating appropriate explorations and actions but doing so in ways aligned with disciplinary interests (e.g., earth and atmospheric scientists in Priority 1, governance experts in Priority 2, economists and engineers in Priority 3, and social and behavioural scientists in Priority 4). From a transdisciplinary perspective, the objective is to create a 'whole is greater than the sum of its parts' approach to understanding the complex, multi-faceted and multi-level problems characteristic of contemporary DRR (Ismail-Zadeh et al., 2017; Paton & Buergelt, 2019). This process starts with creating a superordinate social identity derived from stakeholder involvement in collectively defining and framing (e.g., in relation to the context in which a problem is situated) a problem from a holistic rather than a disciplinary or professional perspective.

One way in which the kind of superordinate social identity required emerges comes from the content of Sections 9.2 and 9.4. Rather than operating under the Priority labels, subsuming the Priorities under a socio-cultural-environmental conceptualization could serve to align disciplinary interests under a socio-cultural-environmental collaborative and cooperative framework for problem formulation and resolution. While it has not been explored for its potential in this regard, the Machizukuri (community-led place-making with care) construct, which seeks to integrate the roles of residents, city planners and emergency managers (Mamula-Seadon, 2018) represents a potential transdisciplinary approach to DRR.

Transdisciplinary methods focus attention on the co-creation and co-implementation of solution-oriented outcomes that represent the holistic contributions of transdisciplinary team members (Ismail-Zadeh et al., 2017). This has other implications; transdisciplinary approaches are consistent with SFDRR Guiding Principles that call for socially inclusive approaches to DRR that result in the cooperative and collaborative efforts of multiple stakeholders taking shared responsibility for DRR outcomes. Transdisciplinary strategies also align well with approaches to DRR governance offered by proponents of adaptive governance (Djalante et al., 2011; Munene et al., 2018), a process well suited to creating the decision-making processes required to accommodate DRR in the increasingly uncertain and dynamic social-ecological systems contexts in which future DRR planning and intervention will inevitably occur (Karpouzoglou et al., 2016).

9.7 Adaptive governance

Proponents of adaptive governance argue that traditional one-size-fits-all approaches to disaster governance are ill-suited to providing the overarching guidance and direction required to effectively manage the diverse circumstances and activities that comprise comprehensive DRR (Djalante et al., 2011; Munene et al., 2018). For example, it is challenging to develop governance systems that can cater for the unique demands that arise when dealing with risk communication versus preparedness (especially when conceptualized as a developmental process) versus recovery versus BBB/capacity development processes, and whose implications change over time and space (Munene et al., 2018).

Governance systems thus face challenges from the need to encompass the implications of evolving patterns of geographic hazard (e.g., from climate change) and population diversity (including how these change over time), and accommodating cross-cultural and developing country issues (Abrams et al., 2015; Djalante et al., 2011; Forino et al., 2015; Karpouzoglou et al., 2016; Munene et al., 2018). To these can be added the need for governance to encompass policies and plans that see governance functioning to ensure that it and the other priorities (understanding disaster risk, preparedness, BBB/capacity development) play complementary and collaborative roles in DRR.

If DRR is to facilitate this kind of coordinated, collaborative and complementary actions between diverse stakeholders (physical and social scientists, governments and societal agencies and functions, NGOs, and private sector entities), adaptive disaster governance processes could represent an appropriate starting point (Abrams et al., 2015; Djalante et al., 2011; Forino et al., 2015; Karpouzoglou et al., 2016; Munene et al., 2018). These points reiterate the value of exploring the importation of transdisciplinary approaches into the DRR toolkit.

A capacity for DRR decision-making in increasingly turbulent times is enabled by adaptive governance accommodating the roles of multiple actors engaged in collaborative, self-organizing relationships involving state, scientific, risk management, private sector and civil society domains to reinforce its utility in these circumstances (Djalante et al., 2011; Karpouzoglou et al., 2016; Mamula-Seadon & McLean, 2015; Munene et al., 2018). These authors argue that adaptive governance is better suited to enabling the development of networks and partnerships involving key stakeholders sharing knowledge and responsibility between actors operating at different levels of governance in ways that support diverse policy and management capabilities and create settings in which innovative learning can occur. However, Karpouzoglou et al. question the utility of adaptive governance as a framework for accommodating the important SFDRR goal of facilitating social justice and managing issues arising from unequal power relations inherent in governance structures and in how they are applied.

Karpouzoglou et al. (2016) called for future research on governance to include work to increase policy (between actors operating at different levels of governance) transparency, develop agency and institutional coordination (e.g., using transdisciplinary approaches), and identify how responsibility for operationalization and oversight is developed, implemented, evaluated and reviewed. Their suggested way forward advocates for renewed emphasis on the role of engagement in governance through participatory, social justice and leadership processes (see Chapter 8), including through informal and voluntary collaboration.

In support of this position, Karpouzoglou et al. (2016) cite Wyborn's (2015) call for directing attention to the relationship between knowledge-making and decision-making and its potential to inform *co-productive governance* in multi-stakeholder settings in ways that integrate setting, knowledge, processes and governance (cf., transdisciplinary operations and transformative learning).

To realize the benefits of this approach, it is important that the agencies involved in co-production are all well prepared not only for fulfilling policy and planning functions but also able to sustain this functionality when disaster strikes. The discussion on this and the preceding sections require the commitment of financial and other resources. The SFDRR Priority 3 called attention to a need for DRR activities to provide a return on investment. The challenges and implications of pursuing are discussed next.

9.8   CBA and evaluation

Chapter 7 discussed the application of CBA processes to DRR. While it has been applied with some success to economic and physical DRR processes, it has been less effective in determining whether environmental and social DRR activities can demonstrate a return on societal and citizen investment (Shreve & Kelman, 2014). In response to this problem, Chapter 7 offered an alternative approach, one based on systematic evaluation, for exploring the effectiveness of social DRR strategies.

Because this is not an area that has attracted much attention, several possible approaches were offered; one based on a developmental approach to preparedness, one using a preparedness drill, another using a theory-based community development evaluation model, and one using QoL to assess the effectiveness of long-term recovery interventions. All demonstrated how process, content and outcome evaluation could be accomplished. It is, however, debatable whether any of these processes would be automatically amenable to CBA. For example, several of the processes in these models derive from people's life experiences and some were developed by applying community development principles. Because community development strategies enable the development of capacities that facilitate action in DRR and everyday life settings, CBA would need to be able to assess outcomes in both daily life and DRR settings. Given the range of circumstances that would need to be considered and the time frames over which assessment would be required, identifying the costs and the benefits would be challenging. It does, however, remain important to be able to demonstrate that DRR strategies are effective. The evaluation approaches described in Chapter 7 offer ways to pursue this goal. However, these remain tentative until additional work is undertaken to explore their utility. There are some other areas that were not covered here but which merit future consideration. One concerns organizational preparedness.

9.9 Organizational continuity planning

While not covered in detail here, Priority 4 also calls for preparedness in government agencies, businesses, and NGOs. A comprehensive discussion on disaster business continuity planning and organizational preparedness can be found in Paton and McClure (2017). Organizational development strategies designed to facilitate agency capacity to create empowering settings may be required to enhance their ability to complement local community recovery and rebuilding initiatives in ways that accommodate procedural and distributive justice in planning and intervention (Guion et al., 2007). The latter has additional implications for accommodating cultural diversity and the cross-cultural application of DRR knowledge in ways that enable international learning and collaboration (see Chapter 5).

9.10 Learning and collaboration in international settings

In Chapter 5, research supporting the ability of CET to predict earthquake and volcanic preparedness in culturally diverse countries illustrated how DRR processes can support international learning and collaboration. Doing so, however, requires the development of intervention practices that integrate culture-general and culture-specific processes (see Figures 5.2, 6.1 and 6.2). The Taiwan work (Figure 5.3) opens the possibility for developing models in Asian (and more collectivistic) countries and testing their cross-cultural equivalence in Western settings. However, before doing so, the functional utility of the model in Figure 5.3 needs to be assessed (e.g., to predict preparedness).

Chapter 5 also called for greater attention to be directed to accommodating multiculturalism and the concomitant need to develop socially just DRR strategies in multicultural countries. The issues canvassed extended beyond cultural diversity per se to include the procedural and distributive justice implications of accommodating, for example, culture distance and the quality of acculturation over time in DRR planning and intervention design. Accommodating these issues in planning starts with understanding how cultural characteristics can affect DRR planning and intervention, and the content of Chapter 5 provides a starting point for this exercise. Another relevant issue here concerns learning from Indigenous peoples.

9.11 DRR in Indigenous populations

While not discussed above, research is providing valuable insights into how Indigenous knowledge can inform learning, and in the case of Australian Indigenous peoples, learning from some 60,000 years of co-existing with hazardous circumstances (Buergelt et al., 2017). Other work is adding a comparative dimension that can facilitate ways of integrating Western and Indigenous knowledge (Mercer et al., 2007). These represent areas of work that must be prioritized in future research.

9.12 Conclusion

In conclusion, Eliot's call that "we shall not cease from exploration" is a vital message for DRR; there remains much to reflect on, discover and apply. While this concluding chapter identified several areas for exploration, many others no doubt remain to be discovered. Future explorations will take place against a backdrop of evolving, uncertain and dynamic environmental threats. This makes it increasingly important that DRR adopts and applies the transdisciplinary, adaptive and transformative processes identified in the preceding chapters to create fertile ground for forging new ways of knowing DRR. The ability of humanity to survive, thrive and prosper requires a commitment not only to continue to explore but to do so in ways that will ensure that we will embark on journeys that continuously culminate in knowing (DRR) for the first time.

References

Abrams, J. B., Knapp, M., Paveglio, T. B., Ellison, A., Moseley, C., Nielsen-Pincus, M., & Carroll, M. C. (2015). Re-envisioning community–wildfire relations in the U.S. West as adaptive governance. *Ecology and Society, 20*(3), 34. http://dx.doi.org/10.5751/ES-07848-200334.

Adams, R. M., Karlin, B., Eisenman, D. P., Blakley, J., & Glik, D. (2017). Who participates in the great ShakeOut? Why audience segmentation is the future of disaster preparedness campaigns. *International Journal of Environmental Research and Public Health, 14*, 1407. https://doi.org/10.3390/ijerph14111407.

Adhikari, M., Paton, D., Johnston, D., Prasanna, R., & McColl, S. T. (2018). Modelling predictors of earthquake preparedness in Nepal. *Procedia Engineering, 212*, 910–17. https://doi.org/10.1016/j.proeng.2018.01.117.

Aldrich, D. P., & Meyer, A. (2015). Social capital and community resilience. *American Behavioral Scientist, 59*, 254–69. https://doi.org/10.1177/0002764214550299.

Andreason, A. R. (2007). Social marketing. In G. T. Gundlach, L. T. Block, & W. L. Wilkie (eds), *Explorations of Marketing in Society* (pp. 664–78). Mason, OH: Thomson/South-Western.

Armaş, I., Cretu, R. Z., & Ionescu, C. (2017). Self-efficacy, stress, and locus of control: the psychology of earthquake risk perception in Bucharest, Romania. *International Journal of Disaster Risk Reduction, 22*, 71–6. https://doi.org/10.1016/j.ijdrr.2017.02.018.

Arneson, E., Deniz, D., Javernick-Will, A., Liel, A., & Dashti, S. (2017). Information deficits and community disaster resilience. *Natural Hazards Review, 18*(4), 04017010. https://doi.org/10.1061/(ASCE)NH.1527-6996.0000251.

Bajek, R., Matsuda, Y., & Okada, N. (2008). Japan's Jishu-bosai-soshiki community activities: analysis of its role in participatory community disaster risk management. *Natural Hazards, 44*, 281–92.

Ballantyne, M., Paton, D., Johnston, D., Kozuch, M., & Daly, M. (2000). *Information on Volcanic and Earthquake Hazards: The Impact on Awareness and Preparation* (GNS Science Report 2000/2). Lower Hut, New Zealand: Institute of Geological and Nuclear Sciences.

Barlow, D. H. (2002). *Anxiety and Its Disorders* (2nd edn). New York: Guilford Press.

Basolo, V., Steinberg, L. J., Burby, R. J., Levine, J., Cruz, A. M., & Huang, C. (2009). The effects of confidence in government and information on perceived and actual preparedness for disasters. *Environment and Behavior, 41*, 338–64. https://doi.org/10.1177/0013916508317222.

Becker, J. S., Paton, D., Johnston, D. M., & Ronan, K. R. (2013). Salient beliefs about earthquake hazards and household preparedness. *Risk Analysis, 33*, 1710–27.

Becker, J. S., Potter, S. H., McBride, S. K., Wein, A., Doyle, E. E. H., & Paton, D. (2019). When the earth doesn't stop shaking: how experiences over time influenced information needs, communication, and interpretation of aftershock information during the Canterbury Earthquake Sequence, New Zealand. *International Journal of Disaster Risk Reduction, 34,* 397–411. https://doi.org/10.1016/j.ijdrr.2018.12.009.

Berkes, F. (2021). *Community-based Conservation.* Cheltenham, UK and Northampton, MA, USA: Edward Elgar Publishing.

Beunen, R., Patterson, J., & Van Assche, K. (2017). Governing for resilience: the role of institutional work. *Current Opinion in Environmental Sustainability, 28,* 10–16.

Bhandari, R., Okada, N., Yokomatsu, M., & Ikeo, H. (2010). Analyzing urban rituals with reference to development of social capital for disaster resilience. *Proceedings IEEE International Conference on Systems, Man and Cybernetics. Istanbul, Turkey.* https://doi.org/10.1109/ICSMC.2010.5642437.

Bočkarjova, M., Van der Veen, A., & Geurts, P. A. T. M. (2009). A PMT-TTM model of protective motivation for flood danger in the Netherlands. ITC Working Papers Series – Paper 3. Enschede, Netherlands: International Institute for Geo-Information Science and Earth Observation.

Buergelt, P. T., & Paton, D. (2014). An ecological risk management and capacity building model. *Human Ecology, 42,* 591–603. https://doi.org/10.1007/s10745-014-9676-2.

Buergelt, P. T., Paton, D., Sithole, B., Sangha, K., Prasadarao, P. S. D. V., Campion, L., & Campion, J. (2017). Living in harmony with our environment: a paradigm shift. In D. Paton & D. M. Johnston (eds), *Disaster Resilience: An Integrated Approach* (2nd edn) (pp. 289–307). Springfield, IL: Charles C. Thomas.

Bulkeley, H. A., & Betsill, M. M. (2005). Rethinking sustainable cities: multilevel governance and the 'urban' politics of climate change. *Environmental Politics, 14*(1), 42–63. https://doi.org/10.1080/0964401042000310178.

Burton, I., Kates, R. W., & White, G. F. (1993). *The Environment as Hazard* (2nd edn). New York: Guilford Press.

Chaiken, S. (1980). Heuristic versus systematic information processing and the use of source versus message cues in persuasion. *Journal of Personality and Social Psychology, 39,* 752–66.

Chaiken, S., & Trope, Y. (1999). *Dual-process Theories in Social Psychology.* New York: Guilford Press.

Charleson, A. W., Cook, B., & Bowering, G. (2003). *Assessing and Increasing the Level of Earthquake Preparedness in Wellington Homes.* Proceedings of the 7th Pacific Conference on Earthquake Engineering, Society for Earthquake Engineering, Wellington, New Zealand.

Charlesworth, M., & Okereke, C. (2010). Policy responses to rapid climate change: an epistemological critique of dominant approaches. *Global Environmental Change, 20,* 121–9. https://doi.org/10.1016/j.gloenvcha.2009.09.001.

Chen, J. C., & Chen, C. Y. (2015). Discourse on Taiwanese forest fires. In D. Paton, P. T. Buergelt, S. McCaffrey, & F. Tedim (eds), *Wildfire Hazards, Risks and Disasters* (pp. 145–66). London: Elsevier.

Cochran, E. S., Vidale, J., & Tanaka, S. (2004). Earth tides can trigger shallow thrust fault earthquakes. *Science*, *306*(5699), 1164–6. https://doi.org/10.1126/science.1103961.

Cohen, S., & Abukhalaf, A. H. I. (2021). Necessity to plan and implement mental health disaster preparedness and intervention plans. *Academia Letters*, Article 3621. https://doi.org/10.20935/AL3507.

Comstock, R. D., & Mallonee, S. (2005). Comparing reactions to two severe tornadoes in one Oklahoma community. *Disasters*, *29*, 277–87. https://doi.org/10.1111/j.0361-3666.2005.00291.x.

Cottrell, A. (2006). Weathering the storm: women's preparedness as a form of resilience to weather hazards in northern Australia. In D. Paton & D. Johnston (eds), *Disaster Resilience: An Integrated Approach* (pp. 128–42). Springfield, IL: Charles C. Thomas.

Crozier, M., McClure, J., Vercoe, J., & Wilson, M. (2006). The effects of land zoning information on judgments about earthquake damage. *Area*, *38*(2), 143–52.

Dalton, J. H., Elias, M. J., & Wandersman, A. (2001). *Community Psychology*. Belmont, CA: Wadsworth.

Daniel, T. C. (2007). Managing individual response: lessons from public health risk behavioral research. In W. E. Martin, C. Raish, & B. Kent (eds), *Wildfire Risk: Human Perceptions and Management Implications* (pp. 103–16). Washington, DC: Resources for the Future.

De Dominicis, S., Fornara, F., Cancellieri, U. G., Twigger-Ross, C., & Bonaiuto, M. (2015). We are at risk, and so what? Place attachment, environmental risk perceptions and preventive coping behaviours. *Journal of Environmental Psychology*, *43*, 66–78.

Dhakal, S. P. (2012). Regional sustainable development and the viability of environmental community organisations: why inter-organisational social capital matters? *Third Sector Review*, *17*(1), 7–28.

DiPasquale, D., & Glaeser, E. L. (1999). Incentives and social capital: are homeowners better citizens? *Journal of Urban Economics*, *45*(2), 354–84.

Djalante, R., Holley, C., & Thomalla, F. (2011). Adaptive governance and managing resilience to natural hazards. *International Journal of Disaster Risk Science*, *2*(4), 1–14. https://doi.org/10.1007/s13753-011-0015-6.

Dooley, D., Catalano, R., Mishra, S., & Serxner, S. (1992). Earthquake preparedness: predictors in a community survey. *Journal of Applied Social Psychology*, *22*, 451–70.

Drury, J. (2012). Collective resilience in mass emergencies and disasters: a social identity model. In J. Jetten, C. Haslam, & S. H. Alexander (eds), *The Social Cure: Identity, Health and Well-being* (pp. 137–50). Hove, England: Psychology Press.

Duval, T. S., & Mulilis, J.-P. (1999). A person-relative-to-event (PrE) approach to negative threat appeals and earthquake preparedness: a field study. *Journal of Applied Social Psychology*, *29*, 495–516. https://doi.org/10.1111/j.1559-1816.1999.tb01398.x.

Earle, T. C. (2004). Thinking aloud about trust: a protocol analysis of trust in risk management. *Risk Analysis*, *24*, 169–83.

Eiser, J. R., Bostrom, A., Burton, I., Johnston, D. M., McClure, J., Paton, D., van der Pligt, J., & White, M. P. (2012). Risk interpretation and action: a conceptual framework for responses to natural hazards. *International Journal of Disaster Risk Reduction*, *1*, 5–16.

Ejeta, L. T., Ardalan, A., Paton, D., & Yaseri, M. (2016). Predictors of community preparedness for flood in Dire-Dawa town, Eastern Ethiopia: applying an adapted version of the Health Belief Model. *International Journal of Disaster Risk Reduction*, *19*, 341–54. http://dx.doi.org/10.1016/j.ijdrr.2016.09.005.

Elliott, J., Haney, T., & Sams-Abiodun, P. (2010). Limits to social capital: comparing network assistance in two New Orleans neighbors devastated by Hurricane Katrina. *Sociological Quarterly*, *51*, 624–48.

Etkin, D. (1999). Risk transference and related trends: driving forces towards more mega-disasters. *Environmental Hazards*, *1*, 69–75.

Faulkner, H., & Ball, D. (2007). Environmental hazards and risk communication. *Environmental Hazards*, *7*, 71–8.

Finucane, M. L., Alhakami, A., Slovic, P., & Johnson, S. M. (2000). The affect heuristic in judgments of risks and benefits. *Journal of Behavioral Decision Making*, *13*(1), 1–17.

Fischhoff, B. (1995). Risk perception and communication unplugged: twenty years of process. *Risk Analysis*, *15*(2), 137–45.

Forino, G., von Meding, J., & Brewer, G. J. (2015). A conceptual governance framework for climate change adaptation and disaster risk reduction integration. *International Journal of Disaster Risk Science*, *6*, 372–84. https://doi.org/10.1007/s13753-015-0076-z.

Frandsen, M., Paton, D., Sakariassen, K., & Killalea, D. (2012). Nurturing community wildfire preparedness from the ground up: evaluating a community engagement initiative. In D. Paton & F. Tedim (eds), *Wildfire and Community: Facilitating Preparedness and Resilience* (pp. 260–80). Springfield, IL: Charles C. Thomas.

Fraser, A., Paterson, S., & Pelling, M. (2016). Developing frameworks to understand disaster causation: from forensic disaster investigation to risk root cause analysis. *Journal of Extreme Events*, *3*, 1650008. https://doi.org/10.1142/S2345737616500081.

Gifford, R., Scannell, L., Kormos, C., Smolova, L., Biel, A., Boncu, S. et al. (2009). Temporal pessimism and spatial optimism in environmental assessments: an 18-nation study. *Journal of Environmental Psychology*, *29*(1), 1–12. https://doi.org/10.1016/j.jenvp.2008.06.001.

Gregg, C., & Houghton, B. (2006). Natural hazards. In D. Paton & D. Johnston (eds), *Disaster Resilience: An Integrated Approach* (pp. 19–39). Springfield, IL: Charles C. Thomas.

Gregg, C., Houghton, B., Paton, D., Swanson, D. A., & Johnston, D. (2004). Community preparedness for lava flows from Mauna Loa and Hualālai volcanoes, Kona, Hawai'i. *Bulletin of Volcanology*, *66*, 531–40.

Gregg, C. E., Houghton, B. F., Paton, D., Swanson, D. A., Lachman, R., & Bonk, W. J. (2008). Hawaiian cultural influences on support for lava flow hazard mitigation measures during the January 1960 eruption of Kilauea volcano, Kapoho, Hawai'i. *Journal of Volcanology and Geothermal Research*, *172*, 300–307.

Grothmann, T., & Reusswig, F. (2006). People at risk of flooding: why some residents take precautionary action while others do not. *Natural Hazards*, *38*, 101–20.
Guion, D. T., Scammon, D. L., & Borders, A. L. (2007). Weathering the storm: a social marketing perspective on disaster preparedness and response with lessons from Hurricane Katrina. *Journal of Public Policy & Marketing*, *26*, 20–32.
Harries, T. (2008). Feeling secure or being secure? Why it can seem better not to protect yourself against a natural hazard. *Health, Risk and Society*, *10*, 479–90. http://dx.doi.org/10.1080/13698570802381162.
Hastings, G., Stead, M., & Webb, J. (2004). Fear appeals in social marketing: strategic and ethical reasons for concern. *Psychology & Marketing*, *21*, 961–86.
Hawkins, R. L., & Maurer, K. (2010). Bonding, bridging and linking: how social capital operated in New Orleans following Hurricane Katrina. *British Journal of Social Work*, *40*, 1777–93.
Healey, P. (2006). Relational complexity and the imaginative power of strategic spatial planning. *European Planning Studies*, *14*(4), 525–46. http://doi.org/10.1080/09654310500421196.
Heller, K., Alexander, D. B., Gatz, M., Knight, B. G., & Rose, T. (2005). Social and personal factors as predictors of earthquake preparation: the role of support provision, negative affect, age, and education. *Journal of Applied Social Psychology*, *35*, 399–422.
Helweg-Larsen, M. (1999). (The lack of) optimistic bias in response to the 1994 Northridge earthquake: the role of personal experience. *Basic and Applied Social Psychology*, *21*, 119–29.
Hofstede, G. (2001). *Culture's Consequences: Comparing Values, Behaviors, Institutions and Organizations across Nations*. Thousand Oaks, CA: Sage.
Hu, S., Yu, M., Que, T., Fan, G., & Xing, H. (2022). Individual willingness to prepare for disasters in a geological hazard risk area: an empirical study based on the protection motivation theory. *Natural Hazards*, *110*, 2087–111. https://doi.org/10.1007/s11069-021-05026-8.
Huang, S.-K., Lindell, M. K., & Prater, C. S. (2016). Who leaves and who stays? A review and statistical meta-analysis of hurricane evacuation studies. *Environment and Behavior*, *48*(8), 991–1029. http://doi.org/10.1177/0013916515578485.
Hutchison, E. D. (2005). The life course perspective: a promising approach for bridging the micro and macro worlds for social workers. *Families in Society*, *86*(1), 143–52. http://doi.org/10.1606/1044-3894.1886.
Irons, M., & Paton, D. (2017). Social media and emergent groups: the impact of high functionality on community resilience. In D. Paton & D. M. Johnston (eds), *Disaster Resilience: An Integrated Approach* (2nd edn) (pp. 194–211). Springfield, IL: Charles C. Thomas.
Ismail-Zadeh, A. T., Cutter, S. L., Takeuchi, K., & Paton, D. (2017). Forging a paradigm shift in disaster science. *Natural Hazards*, *86*, 969–88.
Jang, L., & LaMendola, W. (2006). The Hakka spirit as a predictor of resilience. In D. Paton & D. Johnston (eds), *Disaster Resilience: An Integrated Approach* (pp. 174–89). Springfield, IL: Charles C. Thomas.

Jang, L., & LaMendola, W. (2007). Social work in natural disasters: the case of spirituality and posttraumatic growth. *Advances in Social Work Journal*, 8(2), 67–78.

Jang, L. J., Wang, J. J., Paton, D., & Ning-Yu, T. (2016). Cross-cultural comparisons between the earthquake preparedness models of Taiwan and New Zealand. *Disasters*, 40, 327–45. http://doi.org/10.1111/disa.12144.

Johnson, B. B., & Nakayachi, K. (2017). Examining associations between citizens' beliefs and attitudes about uncertainty and their earthquake risk judgments, preparedness intentions, and mitigation policy support in Japan and the United States. *International Journal of Disaster Risk Reduction*, 22, 37–45. https://doi.org/10.1016/j.ijdrr.2017.02.019.

Johnston, D. M., Bebbington, M., Lai, C.-D., Houghton, B. F., & Paton, D. (1999). Volcanic hazard perceptions: comparative shifts in knowledge and risk. *Disaster Prevention and Management*, 8, 118–26.

Karpouzoglou, T., Dewulf, A., & Clark. J. (2016). Advancing adaptive governance of social-ecological systems through theoretical multiplicity. *Environmental Science & Policy*, 57, 1–9. http://dx.doi.org/10.1016/j.envsci.2015.11.011.

Kerstholt, J., Duijnhoven, H., & Paton, D. (2017). Flooding preparedness in the Netherlands: integrating factors at individual, social and institutional level. *International Journal of Disaster Risk Reduction*, 24, 52–7. http://dx.doi.org/10.1016/j.ijdrr.2017.05.013.

Khan, F., Moench, M., Reed, S. O., Dixit, A., Shrestha, S., & Dixit, K. (2012). *Understanding the Costs and Benefits of Disaster Risk Reduction Under Changing Climate Conditions: Case Study Results and Underlying Principles*. Bangkok: ISET-International.

Kitagawa, K. (2015). Living with an active volcano: informal and community learning for preparedness in south of Japan. *Advances in Volcanology*, 12, 1–17.

Klima, K., & Rueda, I. A. (2020). *Counting the Costs: Improving Disaster Recovery Costs Estimations*. Boulder, CO: Natural Hazards Center.

Kong, L. (2007). Cultural icons and urban development in Asia: economic imperative, national identity, and global city status. *Political Geography*, 26, 383–404. http://doi.org/10.1016/j.polgeo.2006.11.007.

Kull, D., Mechler, R., & Hochrainer-Stigler, S. (2013). Probabilistic cost–benefit analysis of disaster risk management in a development context. *Disasters*, 37, 374–400. https://doi.org/10.1111/disa.12002.

Lau, D., & Murnighan, J. K. (1998). Demographic diversity and faultlines: the compositional dynamics of organizational groups. *Academy of Management Review*, 23, 325–40.

Lau, D., & Murnighan, J. K. (2005). Interactions within groups and subgroups: the dynamic effects of demographic faultlines. *Academy of Management Journal*, 48, 645–50.

Lindell, M. K., & Perry, R. W. (2000). Household adjustment to earthquake hazard: a review of research. *Environment and Behavior*, 32, 461–501.

Lindell, M. K., & Perry, R. W. (2012). The protective action decision model: theoretical modifications and additional evidence. *Risk Analysis*, 32, 616–32. http://dx.doi.org/10.1111/j.1539-6924.2011.01647.x.

Lindell, M. K., Arlikatti, S., & Prater, C. S. (2009). Why people do what they do to protect against earthquake risk: perceptions of hazard adjustment attributes. *Risk Analysis, 29,* 1072–88.

Lion, R., Meertens, R. M., & Bot, I. (2002). Priorities in information desire about unknown risks. *Risk Analysis, 22,* 765–76.

Lopes, R. (1992). *Public Perception of Disaster Preparedness.* Washington, DC: The American Red Cross.

Lupton, D. (1999). *Risk.* London: Routledge.

Mamula-Seadon, L. (2018). Building community resilience through empowerment: place-making in different cultural contexts. In D. Paton, R. Sheng-Her, & L.-J. Jang (eds), *Community-based Disaster Risk Reduction and Recovery: Integrating Community Development and Risk Management* (pp. 50–80). Taipei, Taiwan: Tzu Chi Foundation.

Mamula-Seadon, L., & McLean, I. (2015). Response and early recovery following 4 September 2010 and 22 February 2011 Canterbury earthquakes: societal resilience and the role of governance. *International Journal of Disaster Risk Reduction, 14,* 82–95.

Marti, M., Stauffacher, M., Matthes, J., & Wiemer, S. (2018). Communicating earthquake preparedness: the influence of induced mood, perceived risk, and gain or loss frames on homeowners' attitudes toward general precautionary measures for earthquakes. *Risk Analysis, 38,* 710–23.

Martin, I. M., Bender, H., & Raish, C. (2007). What motivates individuals to protect themselves from risks: the case of wildland fires. *Risk Analysis, 27,* 887–900.

Matsumoto, D., & Juang, L. (2008). *Culture and Psychology.* Belmont, CA: Thompson/Wadsworth.

Mayell, H. (2002). Volcanoes loom as sleeping threat for millions. *National Geographic News.* Retrieved 24 October 2005 from http://www.rense.com.

McAllan, C., McAllan, V., McEntee, K., Gale, W., Taylor, D., Wood, J., Thompson, T., Elder, J., Mutsaers, K., Leeson, W., & Whittlesea, L. C. (2011). *Lessons Learned by Community Recovery Committees of the 2009 Victorian Bushfires.* Melbourne, Australia: Cube Management Solutions.

McBride, S. K., Becker, J. S., & Johnston, D. M. (2019). Exploring the barriers for people taking protective actions during the 2012 and 2015 New Zealand ShakeOut drills. *International Journal of Disaster Risk Reduction, 37,* 1–11. https://doi.org/10.1016/j.ijdrr.2019.101150.

McClure, J. (2017). Fatalism, causal reasoning, and natural hazards. *Oxford Research Encyclopaedia of Natural Hazard Science.* https://doi.org/10.1093/acrefore/9780199389407.013.39.

McClure, J., Allen, M. W., & Walkey, F. (2001). Countering fatalism: causal information in news reports affects judgments about earthquake damage. *Basic and Applied Social Psychology, 23*(2), 109–21. https://doi.org/10.1207/S15324834BASP2302-3.

McClure, J., Sutton, R. M., & Wilson, M. (2007). How information about building design influences causal attributions for earthquake damage. *Asian Journal of Social Psychology, 10,* 233–42.

McClure, J., White, J., & Sibley, C. G. (2009). Framing effects on preparation intentions: distinguishing actions and outcomes. *Disaster Prevention and Management*, 18, 187-99.

McClure, J., Spittal, M. J., Fischer, R., & Charleson, A. (2015). Why do people take fewer damage mitigation actions than survival actions? Other factors outweigh cost. *Natural Hazards Review*, 16(2). https://doi.org/10.1061/(ASCE)NH.1527-6996.0000152.

McIvor, D., & Paton, D. (2007). Preparing for natural hazards: normative and attitudinal influences. *Disaster Prevention and Management: An International Journal*, 16, 79-88.

McKenzie-Mohr, D. (2000). Fostering sustainable behavior through community-based social marketing. *American Psychologist*, May, 531-7.

McKenzie-Mohr, D., & Smith, W. (1999). *Fostering Sustainable Behavior: An Introduction to Community-based Social Marketing*. Gabriola Island, B.C.: New Society.

McLennan, J., Marques, M. D., & Every, D. (2020). Conceptualising and measuring psychological preparedness for disaster: the Psychological Preparedness for Disaster Threat Scale. *Natural Hazards*, 101, 297-307.

McLennan, J., Cowlishaw, S., Paton, D., Beatson, R., & Elliott, G. (2014). Predictors of south-eastern Australian households' strengths of intentions to self-evacuate if a wildfire threatens: two theoretical models. *International Journal of Wildland Fire*, 23, 1176-88. http://dx.doi.org/10.1071/WF13219.

McNamara, K. E., & Buggy, L. (2017). Community-based climate change adaptation: a review of academic literature. *Local Environment*, 22(4), 443-60. http://dx.doi.org/10.1080/13549839.2016.1216954.

Mercer, J., Dominey-Howes, D., Kelman, I., & Lloyd, K. (2007). The potential for combining indigenous and western knowledge in reducing vulnerability to environmental hazards in small island developing states. *Environmental Hazards*, 7, 245-56. https://doi.org/10.1016/j.envhaz.2006.11.001.

Mezirow, J. (2008). An overview on transformative learning. In P. Sutherland & J. Crowther (eds), *Lifelong Learning Concepts and Contexts* (2nd edn) (pp. 24-38). New York: Routledge.

Mileti, D. S., & Darlington, J. D. (1997). The role of searching in shaping reactions to earthquake risk information. *Social Problems*, 44, 89-103.

Mileti, D. S., & O'Brien, P. W. (1993). Public response to aftershock warnings. *US Geological Survey Professional Paper, 1553-B*, 31-42.

Mishra, S., & Suar, D. (2012). Effects of anxiety, disaster education, and resources on disaster preparedness behavior. *Journal of Applied Social Psychology*, 42, 1069-87. https://doi.org/10.1111/j.1559-1816.2011.00853.x.

Monteil, C., Simmons, P., & Hicks, A. (2020). Post-disaster recovery and sociocultural change: rethinking social capital development for the new social fabric. *International Journal of Disaster Risk Reduction*, 42, 101356. https://doi.org/10.1016/j.ijdrr.2019.101356.

Moon, H. C., & Choi, E. K. (2001). Cultural impact on national competitiveness. *Journal of International and Area Studies*, 8(2), 21-36.

Morrissey, S., & Reser, J. (2003). Evaluating the effectiveness of psychological preparedness advice in community cyclone preparedness materials. *Australian Journal of Emergency Management*, 18, 46-61.

Mosel, I., & Levine, S. (2014). *Remaking the Case for Linking Relief, Rehabilitation, and Development*. London: Overseas Development Institute.

Mulilis, J. P., & Duval, T. S. (1995). Negative threat appeals and earthquake preparedness: a person-relative-to-event (PrE) model of coping with threat. *Journal of Applied Social Psychology*, 25, 1319–39.

Mulilis, J. P., Duval, T. S., & Bovalino, K. (2000). Tornado preparedness of students, nonstudent renters, and nonstudent owners: issues of PrE theory. *Journal of Applied Social Psychology*, 30, 1310–29.

Mulilis, J. P., Duval, T. S., & Rogers, R. (2003). The effect of a swarm of local tornados on tornado preparedness: a quasi-comparable cohort investigation. *Journal of Applied Social Psychology*, 33, 1716–25.

Munene, M. B., Swartling, Å. G., & Thomalla, F. (2018). Adaptive governance as a catalyst for transforming the relationship between development and disaster risk through the Sendai Framework. *International Journal of Disaster Risk Reduction*, 28, 653–63.

Nakagawa, Y., & Shaw, R. (2004). Social capital: a missing link to disaster recovery. *International Journal of Mass Emergencies and Disasters*, 22, 5–34.

Ntontis, E., Drury, J., Amlot, R., Rubin, G. J., & Williams, R. (2020). Endurance or decline of emergent groups following a flood disaster: implications for community resilience. *International Journal of Disaster Risk Reduction*, 45, 101493. https://doi.org/10.1016/j.ijdrr.2020.101493.

Nuyen, A. T. (2011). Confucian role-based ethics and strong environmental ethics. *Environmental Values*, 20(4), 549–66. https://doi.org/10.3197/0963 27111X13150367351375.

Onuma, H., Shin, K. J., & Managi, S. (2017). Household preparedness for natural disasters: impact of disaster experience and implications for future disaster risks in Japan. *International Journal of Disaster Risk Reduction*, 21, 148–58. https://doi.org/10.1016/j.ijdrr.2016.11.004.

Papanikolaou, V., Adamis, D., & Kyriopoulos, J. (2012). Long term quality of life after a wildfire disaster in a rural part of Greece. *Open Journal of Psychiatry*, 2, 164–70. https://doi.org/10.4236/ojpsych.2012.22022.

Paton, D. (2007). Measuring and monitoring resilience in Auckland. *GNS Science Report* 2007/18. https://www.massey.ac.nz/massey/fms/Colleges/College%20of%20Humanities%20and%20Social%20Sciences/Psychology/Disasters/pubs/GNS/2007/SR%202007-018%20Auckland%20Resilience.pdf?2D 270087F220B02096617E9EA3B9CAAF.

Paton, D. (2008). Risk communication and natural hazard mitigation: how trust influences its effectiveness. *International Journal of Global Environmental Issues*, 8, 2–16. https://doi.org/10.1504/ijgenvi.2008.017256.

Paton, D. (2013). Disaster resilient communities: developing and testing an all-hazards theory. *Journal of Integrated Disaster Risk Management*, 3, 1–17. https://doi.org/10.5595/idrim.2013.0050.

Paton, D. (2017). Co-existing with natural hazards and their consequences. In D. Paton & D. M. Johnston (eds), *Disaster Resilience: An Integrated Approach* (2nd edn) (pp. 3–17). Springfield, IL: Charles C. Thomas.

Paton, D., & Buergelt, P. T. (2012). Community engagement and wildfire preparedness: the influence of community diversity. In D. Paton & F. Tedim (eds), *Wildfire and Community: Facilitating Preparedness and Resilience* (pp. 241–59). Springfield, IL: Charles C. Thomas.

Paton, D., & Buergelt, P. T. (2019). Risk, transformation and adaptation: ideas for reframing approaches to disaster risk reduction. *International Journal of Environmental Research and Public Health*, 16, 2594. https://doi.org/10.3390/ijerph16142594.

Paton, D., & James, H. (2016). Future directions in integrating recovery and development: theoretical and policy perspective. In H. James & D. Paton (eds), *The Consequences of Disasters: Demographic, Planning, and Policy Implications* (pp. 357–68). Springfield, IL: Charles C. Thomas.

Paton, D., & McClure, J. (2013). *Preparing for Disaster: Building Household and Community Capacity*. Springfield, IL: Charles C. Thomas.

Paton, D., & McClure, J. (2017). Business continuity in disaster contexts. In D. Paton & D. M. Johnston (eds), *Disaster Resilience: An Integrated Approach* (2nd edn) (pp. 79–93). Springfield, IL: Charles C. Thomas.

Paton, D., & Sagala, S. (2018). *Disaster Risk Reduction in Indonesia*. Springfield, IL: Charles C. Thomas.

Paton, D., Buergelt, P. T., & Campbell, A. (2015). Learning to co-exist with environmental hazards: community and societal perspectives and strategies. In J. A. Daniels (ed.), *Advances in Environmental Research* (pp. 1–24). New York: Nova Science Publishers.

Paton, D., Buergelt, P. T., & Prior, T. (2008). Living with bushfire risk: social and environmental influences on preparedness. *Australian Journal of Emergency Management*, 23, 41–8.

Paton, D., Jang, L.-J., & Irons, M. (2015). Building capacity to adapt to the consequences of disasters: linking disaster recovery and disaster risk reduction. In D. Brown (ed.), *Capacity Building: Planning, Programs and Prospects* (pp. 85–114). New York: Nova.

Paton, D., Jang, L.-J., & Liu, L.-W. (2016). Long-term community recovery: lessons from earthquake and typhoon experiences in Taiwan. In H. James & D. Paton (eds), *The Consequences of Disasters: Demographic, Planning, and Policy Implications* (pp. 65–85). Springfield, IL: Charles C. Thomas.

Paton, D., Kerstholt, J., & Skinner, I. (2017). Hazard readiness and resilience. In D. Paton & D. M. Johnston (eds), *Disaster Resilience: An Integrated Approach* (2nd edn) (pp. 114–37). Springfield, IL: Charles C. Thomas.

Paton, D., McClure, J., & Buergelt, P. T. (2006). Natural hazard resilience: the role of individual and household preparedness. In D. Paton & D. Johnston (eds), *Disaster Resilience: An Integrated Approach* (pp. 105–27). Springfield, IL: Charles C. Thomas.

Paton, D., Okada, N., & Sagala, S. (2013). Understanding preparedness for natural hazards: a cross-cultural comparison. *Journal of Integrated Disaster Risk Management*, 3, 18–35.

Paton, D., Smith, L., & Johnston, D. (2005). When good intentions turn bad: promoting natural hazard preparedness. *Australian Journal of Emergency Management*, 20, 25–30.

Paton, D., Anderson, E., Becker, J., & Peterson, J. (2015). Developing a comprehensive model of earthquake preparedness: lessons from the Christchurch

earthquake. *International Journal of Disaster Risk Reduction*, *14*, 37–45. http://dx.doi.org/10.1016/j.ijdrr.2014.11.011.
Paton, D., Buergelt, P., Pavavalung, E., Clarks, K., Jang, L. J., & Kuo, G. (2022). All singing from the same song sheet: DRR and the visual and performing arts. In H. James, R. Shaw, V. Sharma, & A. Lukasiewicz (eds), *Disaster Risk Reduction in Asia Pacific: Governance, Education and Capacity* (pp. 123–45). Sydney: Palgrave.
Paton, D., Jang, L.-J., Kitagawa, K., Mamula-Seadon, L., & Sun, Y. (2017). Coping with and adapting to natural hazard consequences: cross cultural perspectives. In D. Paton & D. M. Johnston (eds), *Disaster Resilience: An Integrated Approach* (2nd edn) (pp. 236–54). Springfield, IL: Charles C. Thomas.
Paton, D., Johnston, D., Mamula-Seadon, L., & Kenney, C. M. (2014). Recovery and development: perspectives from New Zealand and Australia. In N. Kapucu & K. T. Liou (eds), *Disaster & Development: Examining Global Issues and Cases* (pp. 255–72). New York: Springer.
Paton, D., Kelly, G., Buergelt, P. T., & Doherty, M. (2006). Preparing for bushfires: understanding intentions. *Disaster Prevention and Management: An International Journal*, *15*, 566–75. https://doi.org/10.1108/09653560610685893.
Paton, D., Smith, L., Daly, M., & Johnston, D. (2008). Risk perception and volcanic hazard mitigation: individual and social perspectives. *Journal of Volcanology and Geothermal Research*, *172*, 179–88.
Pelling, M. (2011). *Adaptation to Climate Change: From Resilience to Transformation*. Abingdon: Routledge.
Peng, J., Strijker, D., & Wu, Q. (2020). Place identity: how far have we come in exploring its meanings? *Frontiers in Psychology*, *11*(294). https://doi.org/10.3389/fpsyg.2020.00294.
Plous, S. (1993). *The Psychology of Judgment and Decision-making*. New York: McGraw-Hill.
Prior, T., & Paton, D. (2008). Understanding the context: the value of community engagement in bushfire risk communication and education. Observations following the East Coast Tasmania bushfires of December 2006. *Australasian Journal of Disaster and Trauma Studies*, *2*. https://www.massey.ac.nz/~trauma/issues/2008-2/prior.htm.
Quarantelli, E. L. (1996). Ten criteria for evaluating the management of community disasters. Preliminary paper #24 1, Newark, DE: University of Delaware Disaster Research Center.
Rademacher, A. (2015). Urban political ecology. *Annual Review of Anthropology*, *44*, 137–52. http://dx.doi.org/10.1146/annurev-anthro-102214-014208.
Raggio, R. D., & Folse, J. A. G. (2011). Expressions of gratitude in disaster management: an economic, social marketing and public policy perspective on post-Katrina campaigns. *Journal of Public Policy & Management*, *30*, 168–74. http://dx.doi.org/10.1509/jppm.30.2.168.
Ranjbar, M., Soleimani, A. A., Sedghpour, B. S., Shahboulaghi, F. M., Paton, D., & Noroozi, M. (2021). Public intention to prepare for earthquakes: psychometric properties of Persian version. *Iran Journal of Public Health*, *50*, 1678–86.
Reininger, B. M., Rahbar, M. H., Lee, M., Chen, Z., Pope, J., & Adams, B. (2013). Social capital and disaster preparedness among low-income Mexican

Americans in a disaster-prone area. *Social Science & Medicine*, *83*, 50–60. https://doi.org/10.1016/j.socscimed.2013.01.037.
Rippl, S. (2002). Cultural theory and risk perception: a proposal for a better measurement. *Journal of Risk Research*, *5*, 147–65.
Sadeka, S., Mohamad, M. S., Reza, M. I. H., Manap, J., & Sarkar, S. K. (2015). Social capital and disaster preparedness: conceptual framework and linkage. *E-Proceeding of the International Conference on Social Science Research (ICSSR)*, 8 & 9 June 2015, Meliá Hotel Kuala Lumpur, Malaysia.
Sarzynski, A. (2015). Public participation, civic capacity, and climate change adaptation in cities. *Urban Climate*, *14*, 52–67. http://dx.doi.org/10.1016/j.uclim.2015.08.002.
Scheinert, S., & Comfort, L. K. (2014). Finding resilient networks: measuring resilience in post-extreme event reconstruction missions. In N. Kapucu & K. Liou (eds), *Disasters and Development: Examining Global Issues and Cases* (pp. 181–99). New York: Springer.
Seebauer, S., & Babcicky, P. (2017). Trust and the communication of flood risks: comparing the roles of local governments, volunteers in emergency services and neighbors. *Journal of Flood Risk Management*, *11*(3), 305–16. https://doi.org/10.1111/jfr3.12313.
Shane, P., Gehrels, M., Zawalna-Geer, A., Augustinus, P., Lindsay, J., & Chaillou, I. (2013). Longevity of a small shield volcano revealed by crypto-tephra studies (Rangitoto volcano, New Zealand): change in eruptive behavior of a basaltic field. *Journal of Volcanology and Geothermal Research*, *257*, 174–83.
Shreve, C. M., & Kelman, I. (2014). Does mitigation save? Reviewing cost–benefit analyses of disaster risk reduction. *International Journal of Disaster Risk Reduction*, *10*, 213–35. https://doi.org/10.1016/j.ijdrr.2014.08.004.
Siegrist, M., & Cvetkovich, G. (2000). Perception of hazards: the role of social trust and knowledge. *Risk Analysis*, *20*, 713–19.
Siegrist, M., & Gutscher, H. (2008). Natural hazards and motivation for mitigation behavior: people cannot predict the affect evoked by a severe flood. *Risk Analysis*, *28*(3), 771–8.
Silver, A., & Grek-Martin, J. (2015). Now we understand what community really means: reconceptualizing the role of sense of place in the disaster recovery process. *Journal of Environmental Psychology*, *42*, 32–41. https://doi.org/10.1016/j.jenvp.2015.01.004.
Simis, M. J., Madden, H., Cacciatore, M. A., & Yeo, S. K. (2016). The lure of rationality: why does the deficit model persist in science communication? *Public Understanding of Science*, *25*, 400–414. https://doi.org/10.1177/0963662516629749.
Skinner, I. (2016). *Evaluating the Bushfire Ready Neighbourhood (BRN) Program*. Hobart: Decision Support Analytics Pty Ltd/Tasmania Fire Service.
Slotter, R., Trainor, J., Davidson, R., Kruse, J., & Nozick, L. (2020). Homeowner mitigation decision-making: exploring the theory of planned behaviour approach. *Journal of Flood Risk Management*, *13*, e12667. https://doi.org/10.1111/jfr3.12667.
Slovic, P., Fischhoff, B., & Lichtenstein, S. (1982). Facts versus fears: understanding perceived risk. In D. Kahneman, P. Slovic, & A. Tversky (eds), *Judgment*

Under Uncertainty: Heuristic and Biases (pp. 463–92). Cambridge: Cambridge University Press.
Slovic, P., Finucane, M. L., Peters, E., & MacGregor, D. G. (2002). Risk as analysis and risk as feelings: some thoughts about affect, reason, risk, and rationality. Risk Analysis, 24(2), 311–22.
Spialek, M. L., & Houston, J. B. (2019). The influence of citizen disaster communication on perceptions of neighborhood belonging and community resilience. Journal of Applied Communication Research, 47(1), 1–23. https://doi.org/10.1080/00909882.2018.1544718.
Spittal, M. J., McClure, J., Siegert, R. J., & Walkey, F. H. (2005). Optimistic bias in relation to preparedness for earthquakes. Australasian Journal of Disaster and Trauma Studies, 1. http://www.massey.ac.nz/~trauma/issues/2005-1/spittal.htm.
Sullivan, G. B., & Sagala, S. (2020). Quality of life and subjective social status after five years of Mount Sinabung eruptions: disaster management and current sources of inequality in displaced, remaining and relocated communities. International Journal of Disaster Risk Reduction, 49, 101629. https://doi.org/10.1016/j.ijdrr.2020.101629.
Sutton, S. A., Paton, D., Buergelt, P. T., Meilianda, E., & Sagala, S. (2020). What's in a name? 'Smong' and the sustaining of risk communication and DRR behaviours as evocation fades. International Journal of Disaster Risk Reduction, 44, 101408. https://doi.org/10.1016/j.ijdrr.2019.101408.
Sutton, S. A., Paton, D., Buergelt, P. T., Sagala, S., & Meilianda, E. (2020). Sustaining a transformative disaster risk reduction strategy: grandmothers' telling and singing tsunami stories for over 100 years saving lives on Simeulue Island. International Journal of Environmental Research and Public Health, 17. https://doi.org/10.3390/ijerph17217764.
Terpstra, T., & Lindell, M. K. (2013). Citizens' perceptions of flood hazard adjustments: an application of the protective action decision model. Environment and Behavior, 8, 993–1018. https://doi.org/10.1177/0013916512452427.
Thaler, T., & Seebauer, S. (2019). Bottom-up citizen initiatives in natural hazard management: why they appear and what they can do? Environmental Science and Policy, 94, 101–11. https://doi.org/10.1016/j.envsci.2018.12.012.
Twigg, J. (2015). Disaster Risk Reduction. London: Overseas Development Institute.
Uittenbroek, C. J., Mees, H. L. P., Hegger, D. L. T., & Driessen, P. J. (2019). The design of public participation: who participates, when and how? Insights in climate adaptation planning from the Netherlands. Journal of Environmental Planning and Management, 62, 1–19. https://doi.org/10.1080/09640568.2019.1569503.
UNDRR (2015). Sendai Framework for Disaster Risk Reduction 2015–2030. Geneva, Switzerland: United Nations Office for Disaster Risk Reduction.
UNDRR (2016). Report of the Open-Ended Intergovernmental Expert Working Group on Indicators and Terminology Relating to Disaster Risk Reduction. Geneva, Switzerland: United Nations Office for Disaster Risk Reduction. Retrieved from https://www.preventionweb.net/files/50683_oiewgreportenglish.pdf.

UNDRR (2017). *What Is Disaster Risk Reduction?* Retrieved from https://www.UNDRR.org/who-we-are/what-is-drr.
UNDRR (2020). *International Day for Disaster Risk Reduction: The Sendai Seven Campaign in 2020.* Geneva, Switzerland: United Nations Office for Disaster Risk Reduction.
Valenti, M., Masedu, F., Mazza, M., Tiberti, S., Di Giovanni, C., Calvarese, A., Pirro, R., & Sconci, V. (2013). A longitudinal study of quality of life of earthquake survivors in L'Aquila, Italy. *BMC Public Health, 13*, 1143. http://www.biomedcentral.com/1471-2458/13/1143.
Vinnell, L. J., Milfont, T. L., & McClure, J. (2019). Do social norms affect support for earthquake-strengthening legislation? Comparing the effects of descriptive and injunctive norms. *Environment and Behavior, 51*(4), 376–400. https://doi.org/10.1177/0013916517752435.
Vinnell, L. J., Milfont, T. L., & McClure, J. (2020). Why do people prepare for natural hazards? Developing and testing a Theory of Planned Behaviour approach. *Current Research in Ecological and Social Psychology, 2.* https://doi.org/10.1016/j.cresp.2021.100011.
Wachinger, G., Renn, O., Begg, C., & Kuhlcke, C. (2012). The risk perception paradox—implications for governance and communication of natural hazards. *Natural Hazards, 33*, 1049–65. https://doi.org/10.1111/j.1539-6924.2012.01942.x.
Walker-Springett, K., Butler, C., & Adger, W. N. (2017). Wellbeing in the aftermath of floods. *Health and Place, 43*, 66–74. http://dx.doi.org/10.1016/j.healthplace.2016.11.005.
Ward, A. F. (2021). People mistake the internet's knowledge for their own. *Proceedings of the National Academy of Science, 118*(43), e2105061118. https://doi.org/10.1073/pnas.2105061118.
Weinstein, N. D. (1980). Unrealistic optimism about future life events. *Journal of Personality and Social Psychology, 39*, 806–20.
Williams, A., Buergelt, P. T., Schirmer, J., & Paton, D. (2021). 'Enacting some kind of new structure': the potential for collaboration during disaster response and recovery to transform and improve adaptive capacity [ms]. Faculty of Health, University of Canberra.
Witte, K. (1992). Putting the fear back into fear appeals: the extended parallel process model. *Communication Monogram, 59*, 329–49.
Woodgate, G., & Redclift, M. (1998). From a 'sociology of nature' to environmental sociology: beyond social constructionism. *Environmental Values, 7*, 3–24. http://www.jstor.org/stable/30302266.
Wyborn, C. (2015). Co-productive governance: a relational framework for adaptive governance. *Global Environmental Change – Human and Policy Dimensions, 30*, 56–67.
Xiong, D. (2010). Traditional Chinese agricultural culture [Doctoral dissertation, Nanjing Agricultural University].
Yu, J. (2018). A new form of accounting with traditional Chinese cultural thinking through the lens of social and environmental accounting [Doctoral dissertation, The University of Sheffield].

Zaksek, M., & Arvai, J. L. (2004). Toward improved communication about wildfire: mental models research to identify information needs for natural resource management. *Risk Analysis*, *24*(6), 1503-14.

Zimmerman, M. A. (1992). Empowerment theory: psychological, organizational and community levels of analysis. In J. Rappaport & E. Seidman (eds), *The Handbook of Community Psychology* (pp. 43-64). New York: Plenum Press.

Index

Adams, R. M. 106
adaptive capacities 79–80, 81
   in Christchurch 84–5
   in Taiwan 85–7
adaptive demands 80, 87
   collective response to 81, 83
   societal response issues 82
   in Taiwan 82
adaptive governance 145–7
   principles 51–2, 54
   processes 128, 129
Adhikari, M. 52–3
affect-driven decision processes 30
agency–community intervention 52
all-hazards approach 7, 11
Andreason, A. R. 77
anticipation 8, 21, 22–3
   analytical and affect-driven decision processes 30
   anxiety 28–9
   denial and fatalism 27–8
   embarrassment 23–4
   emotion and 30–31
   overconfidence 29–30
   risk and preparedness needs 23
   risk compensation bias 25–7
   socio-environmental perceptions 24
   temporal influences on cost and benefit judgements 31–2
   unrealistic optimism bias 24–5
anxiety 28–9, 95
Arthur, Alan 56–7, 68

Babcicky, P. 120
Bajek, R. 131

BBB construct *see* Build Back Better construct
Becker, J. S. 23
Bhandari, R. 59, 131
Bočkarjova, M. 93–4
BRN strategy *see* Bushfire-Ready Neighbourhoods strategy
Buergelt, P. T. 50–51
Build Back Better construct (BBB construct) 4
Bushfire-Ready Neighbourhoods strategy (BRN strategy) 108–10
Bushfire Relief Centre 123, 133

CA theory *see* Critical Awareness theory
CBA *see* cost–benefit analyses
CET *see* Community Engagement Theory
Chen, C. Y. 12
Chen, J. C. 12
Chonaikai (form of community governance in Japan) 59, 67
city identity, people's sense of 130
climate change 12
Cochran, E. S. 12
co-existence
   DRR 133–4
   socio-environmental 137
collective efficacy 47, 64, 107, 108, 140
collective intelligence resource 118
community
   community-agency preparedness 34, 36
   community-based social marketing strategies 45

community/capacity-building preparedness 34, 35
community–environment level intervention 52
consciousness model 69–71, 123
development
   community development-based approaches 142
   and DRR 140–42
   diversity in recovery processes 76
   participation 47, 64, 107
   preparedness 101
   sense of 49, 59, 78, 107–8, 118, 124
Community Engagement Theory (CET) 46–8, 52–3, 71
   comparative analyses of 71
   cross-cultural applicability assessment 63–8
   research into wildfire preparedness using 108
conflict management strategies 77
Confucius 1–2, 74
content evaluation 92
   countering anticipatory reticence/decide not-to-prepare decisions 95–6
   developmental approach to 93–4, 101
   facilitating comprehensive preparedness 98–100
   facilitating sustained preparedness 102–5
context evaluation 93
coping
   and adaptation constructs to disaster recovery 76–8
   appraisal 44
   competencies 107
co-production approaches 144
co-productive governance 147
cost–benefit analyses (CBA) 9, 90
   and DRR 90–92
   and evaluation 147–8
cost–benefit judgement, temporal influences on 31–2

Critical Awareness theory (CA theory) 45
cross-cultural applicability of CET 63, 72
   comparative research 64–5, 72
   culture-specific characteristics 66
   culture-specific constructs 67
   empirical support for cross-cultural utility 64
   internationally applicable DRR theories 67–8
   relationship between culture-general and culture-specific processes 66–7
cross-cultural DRR 62–3
Crozier, M. 28
cultural beliefs and practices 57–9
cultural dimension model of Hofstede 60–61
cultural diversity
   and collective benefit 60
   and DRR 60–62
cultural orientation 134–5

decision-making 31
denial 27–8, 95
disaster experience 114
   capacity development outcomes 115–16
   emergent learning 115–21
   and inaction 114–15
   repurposing 115, 116–21
   transformative learning 116, 121–9
disaster risk reduction (DRR) 1, 7, 136
   adaptive governance 145–7
   assessing cross-cultural applicability of CET 63–8
   CBA and evaluation 147–8
   community consciousness model 69–71
   community development and 140–42
   cost–benefit analyses and 90–92
   cross-cultural 62–3
   cultural dimensions, place and 131–3

cultural diversity and 60–62
cultural influences on processes and outcomes 57–9
environmental context of 11–12
future preparedness issues 139–40
in indigenous populations 149
in international contexts 56–7
learning and collaboration in international settings 149
need for 1–2
organizational continuity planning 148
risk management strategies 72–3
SFDRR 2–7
socio-cultural-environmental framework 7–8, 10, 137–8
socio-cultural-environmental relationships and 68–9
theories 63
transdisciplinary strategies 144–5
transformative learning 142–4
see also effectiveness assessment of DRR
disaster(s)
governance 61
learning from 113–14
recovery 77–8
risk 10–11
Djalante, R. 51–2
DRR see disaster risk reduction
Drury, J. 77
duration of hazard 15–17

earthquake aftershock sequences 16, 75
earthquake preparedness 64
effectiveness assessment of DRR 89–90
content evaluation 92
applying QoL to evaluating recovery and rebuilding 112
countering anticipatory reticence/decide not-to-prepare decisions 95–6
developmental approach to 93–4, 101
facilitating comprehensive preparedness 98–100
facilitating sustained preparedness 102–5
context evaluation 93
cost–benefit analyses and DRR 90–92
evaluation in recovery and rebuilding settings 110–12
outcome evaluation 92–3
applying QoL to evaluating recovery and rebuilding 112
preparedness drills 101, 106
Shakeout Drill 107
theory-based evaluation 107–10
process evaluation 92
applying QoL to evaluating recovery and rebuilding 112
countering anticipatory reticence/decide not-to-prepare decisions 95–6
developmental approach to 93–4, 101
enabling preparedness 97
facilitating comprehensive preparedness 98–100
facilitating sustained preparedness 102–5
Eliot, T. S. 136, 150
embarrassment 23–4
emergent learning 115–21
emotion and anticipation 30–31
empowerment 48, 64, 108, 140
Epictetus 113
evacuation preparedness 34, 35

fatalism 27–8, 95
fault-lines 50
Folse, J. A. G. 77

gotong royong 67
governance
adaptive 145–7
co-productive 147

preparedness and 51–4
systems 146
and transformative DRR learning 125–9
Gregg, C. E. 13, 57, 58, 72, 91
Group Faultlines theory 49–51

Hakka Spirit 58–9, 60, 86, 87
hazard
behaviours 13, 20–21
characteristics 12–13, 20–21
sources of 8
hazardscape 11–12, 47, 143
HBM *see* Health Belief Model
Health Belief Model (HBM) 43
Hofstede, G. 60–61, 64
Ho-Ping
cultural dimensions 131–2
informative characteristic of local governance 127
role of community leaders in 125
role of local leaders in 126–7
transformative DRR learning outcomes in 128
Houghton, B. F. 91

IC cultural dimension *see* individualism–collectivism cultural dimension
immediate impact 80, 81
Indian Ocean tsunami (2004) 123
Indigenous Hawai'ians, cultural beliefs of 57–8
individualism–collectivism cultural dimension (IC cultural dimension) 61–2, 64, 67
intensity of hazard 15
Irons, M. 118

James, H. 111
Jang, L. J. 58
Jishubo 67
Johnston, D. 23, 114, 115

Kagoshima, Japan
cultural dimensions 131–2

informative characteristic of local governance 127
role of community leaders in 125
role of local leaders in 126–7
transformative DRR learning outcomes in 128
Karpouzoglou, T. 146–7
Kelly, G. 31
Kelman, I. 90–91
Kitagawa, K. 131
Krubutan (reciprocal work commitments) 67
kyojo process 67, 128
Kyozon construct 69, 122, 134

LaMendola, W. 58
large-scale hazard events 22, 23
lava flow hazards 13
learning and collaboration in international settings 149
Linking Relief, Rehabilitation and Development concept 120
livelihood preparedness 34, 36
locational influences in capacity development 130–31
long-term orientation (LTO) 61
Lopes, R. 41
low-cost preparedness 39–42, 55, 96
LTO *see* long-term orientation

Machizukuri (community-led place-making with care) 59, 60, 69
masculinity–femininity dimension (MF dimension) 61
matrix model of person, community and societal preparedness predictors 53–4
McBride, S. K. 23
McClure, J. 148
MF dimension *see* masculinity–femininity dimension
Monteil, C. 120

Nakagawa, Y. 76
natural hazard events 15–16, 43, 142
natural processes 10, 12, 56

cultural beliefs influence people's interpretation of 57
frequency of 13–15
neighbourhood influences in capacity development 130–31
neighbourhood-wide vegetation clearing 50
New South Wales (NSW)
cultural dimensions 132
repurposing 118
role of community leaders in 125
role of local leaders in 126–7
New Zealand
adaptive capacities in 84–5
Christchurch earthquake (2011) 79
cultural dimensions 132
discrepancies between agency functions and community 119–20
emergent capacity development in 116–17
local governance processes in 127
personal, family, community, societal and environmental factors 81
role of community leaders in 125
role of emergent local leaders 127
survivor perspectives 81–2

organizational continuity planning 148
outcome evaluation 92–3
preparedness drills 101, 106
shakeout drill 107
theory-based evaluation 107–10
outcome expectancy beliefs 46
overconfidence of people 29–30, 96

PADM *see* Protective Action Decision Model
Paguyuban (community-based organizations) 67
Paton, D. 14, 22, 26, 31, 32, 40, 50–51, 111, 118, 148
PD dimension *see* power distance dimension
perceived resource availability 44

person–environment level intervention 52
Person-relative-to-Event theory (PrE theory) 44
planning preparedness 34, 35
PMT *see* Protection Motivation Theory
positive outcome expectancy 64
post-disaster capacity development 114, 135
city identity 130
cultural dimensions, place and DRR 131–3
emergent learning 115–21
locational and neighbourhood influences 130–31
repurposing 115, 116–21
socio-environmental relationships and co-existence 133–4
transformative DRR learning 116, 121, 134–5
capacity development 121–3
governance and 125–9
modelling 124–5
power distance dimension (PD dimension) 61–2
pre-contemplation 93
precursory periods 17–18
preparatory reticence of people 113
preparedness 8, 21, 33, 97
community 101
comprehensive 34, 37, 98–100
drills 101, 106
earthquake 64
evacuation 34, 35
future issues 139–40
and governance 51–4
information 37–8
low-cost 39–42, 55, 96
livelihood 34, 36
planning 34, 35
psychological 29, 34–6
and resilience 34, 36–7
role in comprehensive DRR 54–5
structural 34, 35, 36, 40, 42
survival 34, 35, 38–9, 41
sustained 102–5
theories 42–51

types of 34
understanding and facilitating 38–9
PrE theory *see* Person-relative-to-Event theory
Prior, T. 32
process evaluation 92
   countering anticipatory reticence/decide not-to-prepare decisions 95–6
   developmental approach to 93–4, 101
   enabling preparedness 97
   facilitating comprehensive preparedness 98–100
   facilitating sustained preparedness 102–5
protection motivation beliefs 46
Protection Motivation Theory (PMT) 43, 52–3, 71
Protective Action Decision Model (PADM) 45–6
psychological preparedness 29, 34–6

Quality of Life (QoL) 111

Raggio, R. D. 77
recovery and rebuilding settings, evaluation in 110–12
repurposing 115, 116–21
resilience
   collective resilience capability 77
   economic, social, health and cultural 4, 88
   personal and social 37
   preparedness and 34, 36–7
response and recovery settings, DRR in 74–6
   applying coping and adaptation constructs to recovery 76–8
   Christchurch, adaptive capacities in 84–5
   survivor perspectives on response and recovery demands 78–84
   Taiwan, adaptive capacities in 85–7

response demands
   government jobs as 83–4
   survivor perspectives on 78–79, 81–2
response-generated demands 75
response times 17–18
risk
   compensation bias 15, 25–7, 95, 120
   sources of 11, 12, 13
risk information 24–5
   emotional reactions to 29
   people's interpretation of 10
   and preparedness needs 30
   socio-cultural beliefs influencing 13
risk management
   agencies/authorities 26, 47, 49, 50, 64
   implications for 17, 18
   strategies 64, 72–3
   thinking 12
Robinson, Ken 89

Sagala, S. 77, 111
Sambatan (reciprocal assistance between neighbours) 67
Seebauer, S. 120, 127
self-efficacy 44, 107, 140
Sendai Framework for Disaster Risk Reduction (SFDRR) 2, 55, 56, 60, 67, 136
   contribution to DRR 142–3
   goal of facilitating social justice 146–7
   guiding principles 5–6, 48 50, 145
   pivotal ideas 7
   priorities 3–5, 147
   shifts in capacity development by 115
SFDRR *see* Sendai Framework for Disaster Risk Reduction
Shakeout Drill 101, 107
Shane, P. 12
shared responsibility
   in DRR processes 142
   principle 5, 33, 45, 120
   relationships 138

Shaw, R. 76
Shreve, C. M. 90–91
Simeulue, Indonesia
　cultural dimensions 131–2
　informative characteristic of local governance 127
　role of community leaders in 125
　role of local leaders in 126–7
single action bias 39, 139
Smith, L. 26
smong concept 123, 134
social capital 46, 48–9, 108, 140
　bonding 76, 120
　linking 108
　theory 71
social-cognitive biases and beliefs 13, 21
social recovery strategies 77
social support 108
socio-cultural
　diversity 49, 50, 51
　processes 62
　settings 57
socio-cultural-environmental
　framework 7–8, 137–8
　relationships and DRR 68–9
socio-environmental
　perceptions 24
　relationships of DRR 133–4
song lyric analyses 77–8
spatial distribution of hazard consequences 18–19
spirituality 87
Stages of Change or Transtheoretical Model 93–4
structural preparedness 34, 35, 36, 40, 42
Sullivan, G. B. 77, 111
survival preparedness 34, 35, 38–9, 41
survivor perspectives
　adaptive capacities 79–80, 81
　in Christchurch 81–2
　personal, family, community, societal and environmental factors 81, 83
　recovery and rebuilding challenges 79, 80
　on response and recovery demands 78–79, 81–2
　in Taiwan 82–4

Taisho eruption of Sakurajima volcano (1914) 121–2
Taiwan
　adaptive capacities in 85–7
　Chi Chi earthquake (1999) 79
　personal, family, community, societal and environmental factors 83
　repurposing and emergent outcomes 117–18
　survivor perspectives 82–4
Tasmania
　cultural dimensions 132
　local governance processes in 127
　response activities in 118
　role of community leaders in 125
　role of emergent local leaders 127
　wildfire in (2013) 118
temporal distribution 19
Thaler, T. 127
theory-based evaluation 107–10
Theory of Planned Behaviour (TPB) 44–5
thinking global 6
Tōhoku earthquake in Japan (2011) 75
TPB *see* Theory of Planned Behaviour
transdisciplinary strategies 144–5
transformative capacity development process 143
transformative DRR learning 116, 121, 133, 134–5, 142–4
　capacity development 121–3
　governance and 125–9
　modelling 124–5
Tri Hita Karana 69
trust 108, 140

UA dimension *see* uncertainty avoidance dimension
Uittenbroek, C. J. 129
uncertainty avoidance dimension (UA dimension) 61
understanding disaster risk process 10–11

duration of hazard 15–17
environmental context of DRR
    11–12
frequency of natural process
    13–15
hazard behaviours 13, 20–21
hazard characteristics 12–13,
    20–21
intensity of hazard 15
precursory periods 17–18
response times 17–18
spatial distribution 18–19
temporal distribution 19
UNDRR *see* United Nations Office for
    Disaster Risk Reduction
United Nations Office for Disaster
    Risk Reduction (UNDRR) 4,
    22, 43, 66

definition of preparedness 21, 33
goal of enabling enduring
    capacity development
    outcomes 135
goal of supporting international
    collaboration 63
unrealistic optimism bias 24–5, 95

Vinnell, L. J. 101, 106
volcanic events 18, 75

Ward, A. F. 29
whole-of-society approach 7
Williams, A. 76, 118–19, 123, 131, 133
Witte, K. 29
Wyborn, C. 147

Titles in the **Elgar Advanced Introductions** series include:

International Political Economy
Benjamin J. Cohen

The Austrian School of Economics
Randall G. Holcombe

Cultural Economics
Ruth Towse

Law and Development
Michael J. Trebilcock and Mariana Mota Prado

International Humanitarian Law
Robert Kolb

International Trade Law
Michael J. Trebilcock

Post Keynesian Economics
J.E. King

International Intellectual Property
Susy Frankel and Daniel J. Gervais

Public Management and Administration
Christopher Pollitt

Organised Crime
Leslie Holmes

Nationalism
Liah Greenfeld

Social Policy
Daniel Béland and Rianne Mahon

Globalisation
Jonathan Michie

Entrepreneurial Finance
Hans Landström

International Conflict and Security Law
Nigel D. White

Comparative Constitutional Law
Mark Tushnet

International Human Rights Law
Dinah L. Shelton

Entrepreneurship
Robert D. Hisrich

International Tax Law
Reuven S. Avi-Yonah

Public Policy
B. Guy Peters

The Law of International Organizations
Jan Klabbers

International Environmental Law
Ellen Hey

International Sales Law
Clayton P. Gillette

Corporate Venturing
Robert D. Hisrich

Public Choice
Randall G. Holcombe

Private Law
Jan M. Smits

Consumer Behavior Analysis
Gordon Foxall

Behavioral Economics
John F. Tomer

Cost–Benefit Analysis
Robert J. Brent

Environmental Impact Assessment
Angus Morrison Saunders

Comparative Constitutional Law
Second Edition
Mark Tushnet

National Innovation Systems
Cristina Chaminade, Bengt-Åke Lundvall and Shagufta Haneef

Ecological Economics
Matthias Ruth

Private International Law and Procedure
Peter Hay

Freedom of Expression
Mark Tushnet

Law and Globalisation
Jaakko Husa

Regional Innovation Systems
Bjørn T. Asheim, Arne Isaksen and Michaela Trippl

International Political Economy
Second Edition
Benjamin J. Cohen

International Tax Law
Second Edition
Reuven S. Avi-Yonah

Social Innovation
Frank Moulaert and Diana MacCallum

The Creative City
Charles Landry

International Trade Law
Michael J. Trebilcock and Joel Trachtman

European Union Law
Jacques Ziller

Planning Theory
Robert A. Beauregard

Tourism Destination Management
Chris Ryan

International Investment Law
August Reinisch

Sustainable Tourism
David Weaver

Austrian School of Economics
Second Edition
Randall G. Holcombe

U.S. Criminal Procedure
Christopher Slobogin

Platform Economics
Robin Mansell and W. Edward Steinmueller

Public Finance
Vito Tanzi

Feminist Economics
Joyce P. Jacobsen

Human Dignity and Law
James R. May and Erin Daly

Space Law
Frans G. von der Dunk

National Accounting
John M. Hartwick

Legal Research Methods
Ernst Hirsch Ballin

Privacy Law
Megan Richardson

International Human Rights Law
Second Edition
Dinah L. Shelton

Law and Artificial Intelligence
*Woodrow Barfield and
Ugo Pagello*

Politics of International
Human Rights
David P. Forsythe

Community-based Conservation
Fikret Berkes

Global Production Networks
Neil M. Coe

Mental Health Law
Michael L. Perlin

Law and Literature
Peter Goodrich

Creative Industries
John Hartley

Global Administration Law
Sabino Cassese

Housing Studies
William A.V. Clark

Global Sports Law
Stephen F. Ross

Public Policy
B. Guy Peters

Empirical Legal Research
Herbert M. Kritzer

Cities
Peter J. Taylor

Law and Entrepreneurship
Shubha Ghosh

Mobilities
Mimi Sheller

Technology Policy
*Albert N. Link and James A.
Cunningham*

Urban Transport Planning
Kevin J. Krizek and David A. King

Legal Reasoning
*Larry Alexander and Emily
Sherwin*

Sustainable Competitive
Advantage in Sales
Lawrence B. Chonko

Law and Development
Second Edition
*Mariana Mota Prado and Michael
J. Trebilcock*

Law and Renewable Energy
Joel B. Eisen

Experience Economy
Jon Sundbo

Marxism and Human Geography
Kevin R. Cox

Maritime Law
Paul Todd

American Foreign Policy
Loch K. Johnson

Water Politics
Ken Conca

Business Ethics
John Hooker

Employee Engagement
Alan M. Saks and Jamie A. Gruman

Governance
Jon Pierre and B. Guy Peters

Demography
Wolfgang Lutz

Environmental Compliance and Enforcement
LeRoy C. Paddock

Migration Studies
Ronald Skeldon

Landmark Criminal Cases
George P. Fletcher

Comparative Legal Methods
Pier Giuseppe Monateri

U.S. Environmental Law
E. Donald Elliott and Daniel C. Esty

Gentrification
Chris Hamnett

Family Policy
Chiara Saraceno

Law and Psychology
Tom R. Tyler

Advertising
Patrick De Pelsmacker

New Institutional Economics
Claude Ménard and Mary M. Shirley

The Sociology of Sport
Eric Anderson and Rory Magrath

The Sociology of Peace Processes
John D. Brewer

Social Protection
James Midgley

Corporate Finance
James A. Brickley and Clifford W. Smith Jr

U.S. Federal Securities Law
Thomas Lee Hazen

Cybersecurity Law
David P. Fidler

The Sociology of Work
Amy S. Wharton

Marketing Strategy
George S. Day

Scenario Planning
Paul Schoemaker

Financial Inclusion
Robert Lensink, Calumn Hamilton and Charles Adjasi

Children's Rights
Wouter Vandenhole and Gamze Erdem Türkelli

Sustainable Careers
Jeffrey H. Greenhaus and Gerard A. Callanan

Business and Human Rights
Peter T. Muchlinski

Spatial Statistics
Daniel A. Griffith and Bin Li

The Sociology of the Self
Shanyang Zhao

Artificial Intelligence in Healthcare
Tom Davenport, John Glaser and Elizabeth Gardner

Central Banks and Monetary Policy
Jakob de Haan and Christiaan Pattipeilohy

Megaprojects
Nathalie Drouin and Rodney Turner

Social Capital
Karen S. Cook

Elections and Voting
Ian McAllister

Negotiation
Leigh Thompson and Cynthia S. Wang

Youth Studies
Howard Williamson and James E. Côté

Private Equity
Paul A. Gompers and Steven N. Kaplan

Digital Marketing
Utpal Dholakia

Water Economics and Policy
Ariel Dinar

Disaster Risk Reduction
Douglas Paton